Catalysis

Volume 13

A Specialist Periodical Report

Catalysis
Volume 13

A Review of Recent Literature

Senior Reporter

James J. Spivey, *Research Triangle Institute, Research Triangle Park, North Carolina, USA*

Reporters

C. Cristiani, *Politecnico di Milano, Milano, Italy*

P. Forzatti, *Politecnico di Milano, Milano, Italy*

G. Groppi, *Politecnico di Milano, Milano, Italy*

Yoshihiro Kubota, *Gifu University, Gifu, Japan*

Ian S. Metcalfe, *Imperial College of Science, Technology and Medicine, London, UK*

Georgios Papadogianakis, *Delft University of Technology, Delft, The Netherlands*

Roger A. Sheldon, *Delft University of Technology, Delft, The Netherlands*

Yoshihiro Sugi, *Gifu University, Gifu, Japan*

Israel E. Wachs, *Lehigh University, Bethlehem, Pennsylvania, USA*

THE ROYAL SOCIETY OF CHEMISTRY
Information Services

ISSN 0140–0568
ISBN 0-85404-209-1

Published by The Royal Society of Chemistry,
Thomas Graham House, Science Park, Milton Road, Cambridge CB4 4WF, UK

Typeset by Computape (Pickering) Ltd, Pickering, North Yorkshire, UK
Printed by Athenaeum Press Ltd, Gateshead, Tyne & Wear

Preface

I have appreciated the many positive comments I have received from you this past year about the Catalysis series. Volume 13 represents an effort to continue to bring to you topics that are both interesting, and important.

A recent paper by R.M. Lambert of Cambridge reminded me of the importance of catalysis to our daily lives, and to the economies of the world (R.M. Lambert, Introduction to Heterogeneous Catalysis, NATO ASI Series E, Vol. 331, 1997, p. 1-26). More than 20% of the entire world's GNP is derived from heterogeneously catalysed processes, corresponding to $5 trillion per year. More than 90% of the output of the world's chemical industry depends on these same types of processes. I expect that this is more than you would have guessed; it is certainly more than I did. I hope that this volume helps our understanding of this important area of work in which we are involved.

This volume consists of five chapters. The first is by Dr. Ian Metcalfe of Imperial College, and deals with solid electrolyte cells for catalyst screening. These materials are important as sensors and control monitors the solid catalysts we all use.

Dr. Israel Wachs of Lehigh University discusses molecular engineering on oxide catalysts. These materials are especially important for partial oxidation reactions, in which selectivity is difficult to control. This chapter focuses on vanadium-based catalysts, and the approach is applicable to other supported catalysts as well.

Drs. Yoshihiro Sugi and Yoshihiro Kubota next present a review of the zeolite-catalysed alkylation of polynuclear aromatics. Their chapter gives an especially useful and general approach to understanding deactivation in these materials.

Drs. Gianpierro Groppi, C. Cristiani, and P. Forzatti discuss the preparation of hexaaluminate materials for high temperature catalytic combustion, where temperatures are typically over 1400 °C. They discuss these unusually stable materials, how they can be modified for specific needs, and how they can be prepared in a simple way.

Finally, Drs. Georgios Papadogianakis and Roger Sheldon review catalysis in aqueous systems. These types of reactions are especially important in a new class of environmentally benign synthesis processes that are being developed in industry.

I want to thank the authors for the uniformly high quality of their manuscripts and their attention to the reviewers' comments. I especially appreciate the patience of Dr. Metcalfe, whose manuscript was submitted very early in this publication cycle. I also appreciate the work of Sharon Davis and Jan Shirley of RTI's Publication and Design Services. They re-drew figures and provided a thorough and professional editing of the final manuscripts. I look forward to working with them on future volumes.

James J. Spivey
Research Triangle Institute
Research Triangle Park, NC USA
email "jjs@rti.org"

Contents

1
Solid Electrolyte Electrochemical Cells for Catalyst Sensing

BY IAN S. METCALFE

1 Introduction

The development of sensors for industrial process monitoring and control is an area of increasing importance. In particular, there are relatively few sensors that are capable of monitoring the state of a catalyst despite the fact that catalyst state can have a very significant impact on overall process performance. Consequently, there is a need to develop new sensors for the in-situ monitoring of catalyst state. Solid electrolyte electrochemical cells show promise as sensors which could be used for intermediate and high temperature application (temperatures greater than about 200°C).

A solid electrolyte is a material in which the electrolytic, or ionic, conductivity is much greater than the electronic conductivity (for solid electrolytes to be practically useful the ratio of electrolytic to electronic conductivities should be of the order of 100 or greater[1,2]). Solid electrolytes with conduction ions of O^{2-}, H^+, Li^+, Na^+, Ag^+, F^-, Cl^- have all been reported. Much attention has been devoted to oxygen-ion conducting solid electrolytes, many of which show appreciable oxygen-ion conductivities in the range of 200-1200°C.

At high oxygen partial pressures an equilibrium is established between the gas phase oxygen, interstitial oxygen ions and electron holes and conductivity is predominantly p-type due to electron-hole transfer.

$$O_2(g) \Leftrightarrow 2O_i'' + 4h^{\cdot} \tag{1.1}$$

Using the notation of Kröger and Vink,[3] the symbol O_i'' denotes an interstitial oxygen ion carrying an effective double negative charge and $h^{\cdot}$ denotes an electron hole.

At low oxygen partial pressures an equilibrium is established between lattice oxygen, gas phase oxygen, oxygen ion vacancies and free electrons and conduction is n-type.

$$2O_O \Leftrightarrow O_2 + V_O^{\cdot\cdot} + 2e' \tag{1.2}$$

O_O denotes a lattice oxygen, $V_O^{\cdot\cdot}$ denotes an oxygen ion vacancy with an effective double positive charge and e′ denotes an electron.

Over a large range partial pressures of oxygen ionic conductivity dominates and the material behaves as a solid electrolyte. Under these conditions there is an equilibrium established between oxygen ion vacancies, interstitial oxygen ions and lattice oxygen.

$$2O_O \Leftrightarrow V_O^{\cdot\cdot} + O_i'' \qquad (1.3)$$

Hence the partial pressure of oxygen and the temperature determine whether the solid will exhibit n-type, p-type or ionic conduction. Although the concentration of defects is important it is also necessary to consider the mobilities of the individual defects; higher ionic mobilities will result in a larger domain for electrolytic conduction. Figure 1[4] shows the dominant mode of conduction in some mixed oxide materials, exhibiting solid electrolyte behaviour, as a function of temperature and oxygen partial pressure.

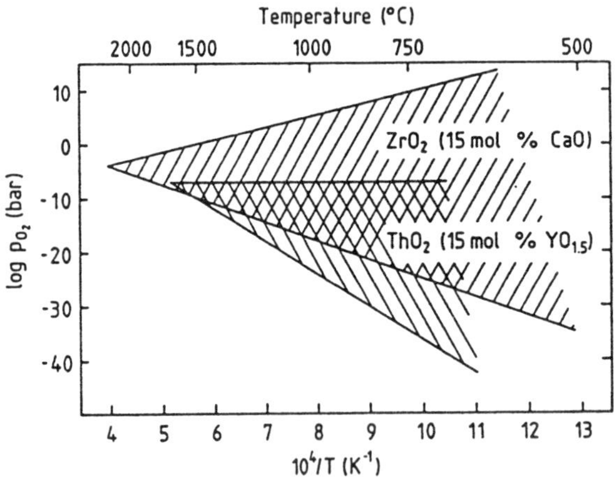

Figure 1 *Range of temperatures and oxygen partial pressures in which the electronic conductivity of solid electrolytes amounts to less than 1% of the total conductivity.*
(Reproduced from *J. Appl. Electrochem.*, 1986, **16**, 475, by permission from Chapman & Hall Ltd.)

Suitable solid electrolytes can be employed as the electrolyte in an electrochemical cell. The electrolyte is used in the form of a membrane which is impermeable to gas phase transport. Electroactive materials, or electrodes, are deposited on both sides of the electrolyte to increase the rates of charge transfer across the electrolyte interface and it is important that the active molecules in the gas phase have easy access to the electrode/electrolyte interface where they can participate in the charge-transfer reactions. For this reason it is necessary, in most cases, to ensure that the electrode has a high porosity while, at the same time, remaining electrically continuous.

Solid electrolyte electrochemical cells can be operated in a variety of ways (the three modes of operation are illustrated schematically in Figure 2). Such a cell may be operated potentiometrically in order to investigate the behaviour of a catalyst of interest. This technique has become known as solid electrolyte potentiometry (SEP). The catalyst itself is deposited in the form of an electrode

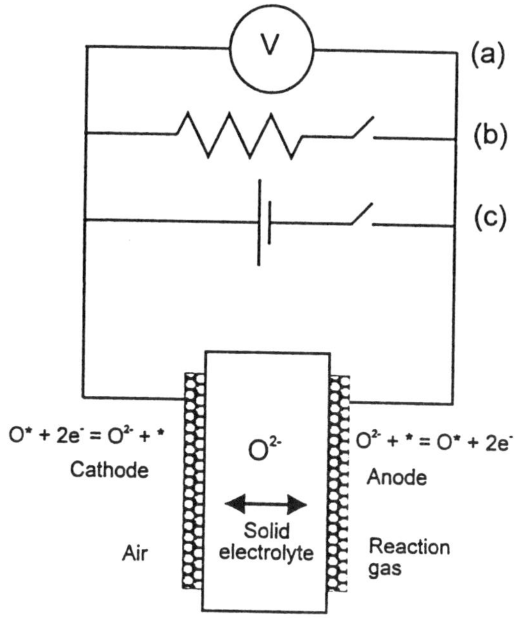

Figure 2 *Schematic showing (a) potentiometric operation, (b) fuel cell operation and (c) electrochemical oxygen pumping.*

and is exposed to reaction conditions.* The electrode on the other side of the solid electrolyte membrane is exposed to constant gas phase conditions and acts as a reference. The e.m.f. generated by such a cell is related to the state of the catalyst and can be indicative of the catalyst work function.[5] In the case of an oxygen-ion conducting solid electrolyte, the e.m.f. of the cell is a reflection of the thermodynamic activity of oxygen at the catalyst-electrode.

If an external circuit and a power supply are now used, a current can be passed through the cell resulting in a pumping of oxygen ions (in the case of an oxygen-ion conducting solid electrolyte) towards the electrode of interest or, if the current is reversed, a removal of oxygen ions from the electrode. Consequently, the surface of the catalyst-electrode can be investigated by the passage of current (amperometric techniques). Recently behaviour under amperometric conditions has been recognised to be more complex because of evidence that electrochemical oxygen pumping (EOP) may modify the behaviour of a catalyst.[5]

* Electrode preparation can be a significant problem for some catalyst systems. In general, it is necessary to sinter the catalyst to ensure adherence to the electrolyte substrate. This does not present much of a problem for metal electrodes but for oxides, where a particular phase may be required, the need for sintering can cause difficulties. This review will not, however, deal with the details of electrode preparation; rather the reader should refer to an original article for details of preparation of a particular electrode (a short review of electrode preparation in solid electrolyte electrochemical cells can be found in reference 5).

Solid electrolyte electrochemical cells can also be operated as fuel cells. One electrode is exposed to a fuel while the other is exposed to an oxidant. The driving force due to the differences in oxygen chemical potential at the two electrodes causes electrons to flow around an external circuit and electrical power is generated. The temperature of operation of fuel cells must be greater because of the increased oxygen fluxes required. Fuel cells are commonly operated at temperatures in excess of 800°C. Fuel cells can also be used for the cogeneration of both useful chemicals and electrical energy if the reaction occurring at the anode is appropriate. Fuel cell technology as such is, however, not relevant for sensor development.

It is important to appreciate that solid electrolyte systems are different from aqueous systems in one important way. In a solid electrolyte system the charge transfer takes place across the catalyst/electrolyte interface whereas the chemical reaction takes place on the gas-exposed surface of the catalyst. In aqueous systems the surface at which charge-transfer occurs is the same as the surface over which any catalysis occurs. As a result care must be exercised when making analogies between the two types of cell.

A number of reviews have been written on the use of solid electrolyte electrochemical cells in catalysis. Vayenas and other workers have reviewed the area of SEP[6,7] as have Gellings et al.,[8] Stoukides[9] and Metcalfe.[10] Vayenas et al.[5,11,12] have also reviewed work on the modification of catalytic behaviour.

The aim of this review is to present and discuss recent work on solid electrolyte electrochemical cells relevant to in-situ catalyst sensing. Consequently, the area of SEP will be concentrated upon, however, appropriate closed-circuit or amperometric studies will also be discussed. This review is intended to also introduce the reader familiar with heterogeneous catalysis to the electrochemical concepts and techniques required to fully appreciate the research work in this field.

Much work with solid electrolyte electrochemical cells has involved the use of oxygen-ion conductors and it is therefore the use of these conductors which will be concentrated upon in this review article.

2 Electrochemical Principles

In this section both the open-circuit and closed circuit behaviour of electrochemical cells will be briefly discussed. The mechanism of the charge-transfer process for oxygen-ion conducting systems will also be discussed.

2.1 Open-circuit Behaviour

2.1.1 Single Charge-transfer Process – Consider an electrochemical cell consisting of two porous platinum electrodes contacting an oxygen-ion conducting solid electrolyte membrane. The two electrodes are exposed to two different chemical potentials of oxygen μ_{O2}' and μ_{O2}''. The cell may be represented as,

$$\mu_{O2}', \mathrm{Pt} \mid \mathrm{MO} \mid \mathrm{Pt}, \mu_{O2}'' \tag{2.1}$$

where MO represents the metal oxide electrolyte.

To determine the e.m.f. of the cell the idea of electrochemical potential will be introduced. The electrochemical potential of species i is defined as,

$$\bar{\mu}_i = \mu_i + z_i F\phi \tag{2.2}$$

where ϕ is the electrostatic potential, z_i is the charge on the species and F is the faraday constant.

Under open-circuit conditions there can be no net flow of current through the cell. However, there may be individual currents due to the migration of each charge carrying species, given by,

$$I_i = -\left(\frac{\sigma_i}{z_i F}\right)\frac{d\bar{\mu}_i}{dx} \tag{2.3}$$

where x is the distance co-ordinate and σ_i is the individual conductivity. The net current must be zero, so,

$$\sum_i I_i = 0 \tag{2.4}$$

The charge-transfer reaction† at both electrodes is assumed to be,

$$O_2 + 4e^- \Leftrightarrow 2O^{2-} \tag{2.5}$$

Using equations (2.2) to (2.4) it can be easily shown that the e.m.f. of the cell will be given by,[13]

$$E_{cell} = \frac{1}{4F}\int_{\mu'_{O_2}}^{\mu''_{O_2}} t_{ion}\, d\mu_{O_2} \tag{2.6}$$

where t_{ion} is the ionic transport number (equal to unity if there is no electronic conduction). If t_{ion} is less than unity the relationship between t_{ion} and μ_{O2} must be known in order to integrate equation (2.6). In practice, if any appreciable electronic conduction is present, then electrons will migrate in the opposite direction to oxygen ions and, if diffusional processes are too slow, the result may be the build-up of oxide at the negative electrode and the appearance of metal at the positive electrode. In such a case, the oxygen potentials at the interfaces are no longer fixed and the cell will exhibit an unstable e.m.f. For solid electrolytes, as t_{ion} approaches unity, this is not a problem and the equation can be easily integrated,

$$E_{cell} = \frac{1}{4F}\left(\mu''_{O_2} - \mu'_{O_2}\right) \tag{2.7}$$

† The nature and mechanism of the charge-transfer reaction will be discussed in more detail in the Section 2.3.

If the oxygen at the interfaces is in equilibrium with the gas phase oxygen, i.e. there is no chemical reaction, and oxygen behaves as an ideal gas, this reduces to,

$$E_{cell} = \frac{RT}{4F} \ln \frac{P''_{O_2}}{P'_{O_2}} \tag{2.8}$$

Equation (2.8) is, of course, the Nernst equation. From the e.m.f. of the cell, an unknown oxygen partial pressure can be determined. This has led to the use of solid electrolyte electrochemical cells as oxygen probes.[4,14]

2.1.2 Mixed Potentials – When reactive gases other than oxygen are also present, it is possible for mixed potentials to occur. As an example, if a mixture of oxygen, carbon monoxide and carbon dioxide are now supplied to the cell,

$$O_2, CO, CO_2, Pt \mid MO \mid Pt, O_2 \tag{2.9}$$

there may be more than one charge-transfer process which can take place, e.g., both of the following reactions may proceed,

$$* + O_O \Leftrightarrow O* + V_O^{\cdot\cdot} + 2e' \tag{2.10}$$

$$CO* + O_O \Leftrightarrow CO_2* + V_O^{\cdot\cdot} + 2e' \tag{2.11}$$

where * denotes an electrode site.

Under open-circuit conditions no net current can flow so that the total rate of the anodic reactions must equal that of the cathodic reactions. Applying this condition allows the e.m.f. to be determined if the current-voltage relationships are known for the charge-transfer processes.[15]

Mixed potentials have been found to be important in reaction systems other than CO oxidation. Michaels and co-workers[16] studied a platinum electrode exposed to nitric oxide, nitrogen dioxide and oxygen in the region of 600-800°C. Using e.m.f. data and closed-circuit work at low overpotentials (to avoid modifying coverages of electroactive species) it was shown that two charge-transfer reactions were important,

$$* + O_O \Leftrightarrow O* + V_O^{\cdot\cdot} + 2e' \tag{2.12}$$

$$NO* + O_O \Leftrightarrow NO_2* + V_O^{\cdot\cdot} + 2e' \tag{2.13}$$

Mixed potentials have also been shown for the electrodes exposed to reacting mixtures of methane and oxygen.[17]

2.2 Closed-circuit Behaviour – When a voltage is applied to such a cell there is a tendency for oxygen to migrate from one electrode to the other. As a result such arrangements have been used as oxygen pumps.[18-20]

When the potential difference across the electrode/electrolyte double-layer is not at its equilibrium value then the interface is said to be polarised. When the interface is polarised a net current will flow, the magnitude of the current being dependent upon the difference in the total anodic and total cathodic currents,

$$i = i_A \text{-} i_C \tag{2.14}$$

Take as an example the oxygen charge-transfer reaction

$$* + O_O \Leftrightarrow O^* + V_O^{\cdot\cdot} + 2e' \tag{2.15}$$

Oxygen from the electrolyte reacts with a vacant site on the electrode forming adsorbed oxygen and creating an oxygen vacancy. The reverse reaction involves oxygen adsorbed on the electrode undergoing electroreduction. The forward reaction is the anodic process and the reverse reaction is the cathodic process. Using transition state theory, anodic and cathodic current densities can be expressed in terms of the potential difference across the interface, the coverage of reactant and the relevant rate constant,

$$i_A = 2Fk_A\theta_* \exp\left[(1-\beta)\frac{F\Delta\phi}{RT}\right] \tag{2.16}$$

$$i_C = 2Fk_C\theta_O \exp\left[-2\beta\frac{F\Delta\phi}{RT}\right] \tag{2.17}$$

where i is the current density, k is the rate constant, the subscript 'A' refers to the anodic process, the subscript 'C' refers to the cathodic process, θ refers to fractional coverages on the electrode, β is the symmetry factor, $\Delta\phi$ is the potential difference across the electrode/electrolyte interface. At equilibrium, the anodic and cathodic current densities are equal in magnitude and equal to what is called the exchange current density, i_0.

$$i_A = i_C = i_0 = 2Fk_A\theta_* \exp\left[2(1-\beta)\frac{F\Delta\phi^0}{RT}\right] = 2Fk_C\,\theta_O \exp\left[-2\beta\frac{F\Delta\phi^0}{RT}\right] \tag{2.18}$$

Therefore, rearranging, the exchange current density can be expressed in terms of the rate constants and coverages,

$$i_0 = 2Fk_C\theta_O\left(\frac{k_C\theta_O}{k_A\theta_*}\right)^{-\beta} \tag{2.19}$$

The actual potential difference across the electrode/electrolyte interface minus the equilibrium potential difference across the interface is known as the electrode overpotential, η, and is, in effect, the driving force for net charge-transfer,

$$\eta = \Delta\phi - \Delta\phi^0 \tag{2.20}$$

Equations (2.16) and (2.17) can be rewritten in terms of the exchange current density and the overpotential,

$$i_A = i_0 \exp\left[2(1-\beta)\frac{\eta F}{RT}\right] \tag{2.21}$$

$$i_C = i_0 \exp\left[-2\beta\frac{\eta F}{RT}\right] \tag{2.22}$$

The net current can be written as,

$$i = i_0\left\{\exp\left[2(1-\beta)\frac{\eta F}{RT}\right] - \exp\left[-2\beta\frac{\eta F}{RT}\right]\right\} \tag{2.23}$$

which is an equation of the Butler-Volmer type.

When,

$$\left|\frac{\eta F}{RT}\right| << 1 \tag{2.24}$$

it is possible to make a low-field approximation for the Butler-Volmer equation,

$$i = 2i_0\frac{\eta F}{RT} \tag{2.25}$$

and there is a linear relationship between the current density and the over-potential.

When,

$$\frac{\eta F}{RT} >> 1, or, \frac{\eta F}{RT} << -1 \tag{2.26}$$

it is possible to make a high-field approximation in a form known as the Tafel equation,

$$i = i_0 \exp\left[2(1-\beta)\frac{\eta F}{RT}\right], or, i = -i_0 \exp\left[-2\beta\frac{\eta F}{RT}\right] \tag{2.27}$$

So far it has been assumed that charge-transfer is the limiting step in the overall process. However, other steps may influence the overall rate. Such steps would include any chemical reactions, adsorption and desorption, and diffusional resistances in the gas phase, on the electrode surface, or in the electrolyte lattice. When mass transfer (here mass transfer refers to any step other than the charge-transfer reaction) between the surface of the electrode and the gas phase is rate limiting, the surface coverages of adsorbed species may be modified by the passage of a current. Under these conditions the current-overpotential relationship is modified. It can easily be shown (assuming that the rate of mass transfer between the three-phase region and the gas phase is directly proportional to the effective oxygen concentration difference between these two regions - a complete derivation has been presented by Bard and Faulkner[21]) that,

$$i = i_0 \left\{ \left(1 - \frac{i}{i_{lA}}\right) \exp\left[2(1-\beta)\frac{\eta F}{RT}\right] - \left(1 - \frac{i}{i_{lC}}\right) \exp\left[-2\beta\frac{\eta F}{RT}\right] \right\} \tag{2.28}$$

where i_{lA} and i_{lC} are the anodic and cathodic limiting currents.* Again this equation can be linearized for low overpotentials,

$$\eta = \frac{RT}{2F}\left(\frac{1}{|i_0|} + \frac{1}{|i_{lC}|} + \frac{1}{|i_{lA}|}\right) i \tag{2.29}$$

Defining resistances,

$$R_{mt} = \frac{RT}{2F}\left(\frac{1}{|i_l|}\right) \tag{2.30}$$

$$R_{ct} = \frac{RT}{2F}\left(\frac{1}{|i_0|}\right) \tag{2.31}$$

where R_{mt} can be considered a resistance due to mass transfer and R_{ct} can be considered as a resistance due to charge transfer. Equation (2.29) then becomes,

$$\eta = i(R_{ct} + R_{mt,A} + R_{mt,C}) \tag{2.32}$$

and the interface can be modelled as a series of resistances.

If a cell is operated potentiometrically, then only two electrodes are required. However, if closed-circuit operation is to be employed it is necessary to measure electrode overpotentials. To measure electrode overpotentials it is necessary to use a three-electrode DC system. The system consists of a working electrode which is exposed to reaction conditions, a counter electrode (which is exposed to a constant partial pressure of oxygen in the case of an oxygen-ion conducting electrolyte), and a reference electrode. The reference electrode remains unpolarized during the measurements and may be exposed to either reaction conditions or the constant conditions of the counter electrode. A power supply is used to polarise the working and counter electrodes while the current is recorded (see Figure 3).

At steady state (once the electrodes have polarised) the applied voltage can be related to the current in the following manner,

$$V_{WC} = E_{WC} + \eta_W(I) + \eta_C(I) + (IR)_{WC} \tag{2.33}$$

where η_W and η_C are the overpotentials at the working and counter electrodes, respectively, and $(IR)_{WC}$ is the resistive loss between the working and counter

* Anodic and cathodic limiting currents are exhibited because of the rate limiting mass transfer between the electrode and the gas phase. As the potential of the electrode is changed the rate of mass transfer will remain constant and, hence, this behaviour will be manifested as a plateau in the current-overpotential relationship.

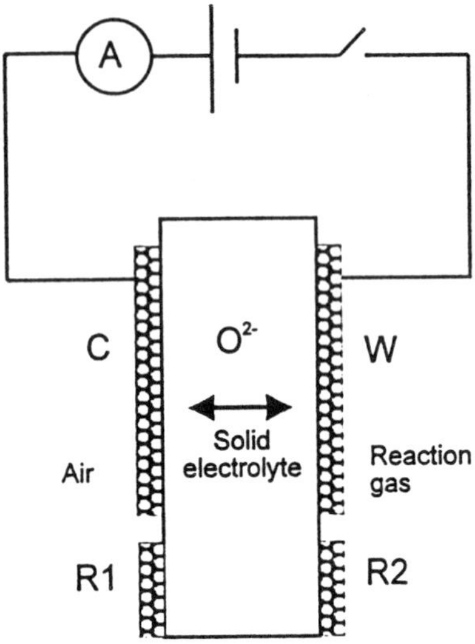

Figure 3 *Schematic showing three-electrode DC operation. C refers to the counter electrode, W refers to the working electrode. The reference electrode can either be on the counter side (R1) or the working side (R2).*

electrodes. The potential difference between the reference electrode and the counter electrode can also be measured and is related to the current in a similar way,

$$V_{RC} = E_{RC} + \eta_C(I) + (IR)_{RC} \tag{2.34}$$

If the potential difference V_{RC} is subtracted from V_{WC},

$$V_{WC} - V_{RC} = E_{WC} - E_{RC} + \eta_W(I) + (IR)_{WC} - (IR)_{RC} \tag{2.35}$$

then the overpotential at the working electrode can be separated from that at the counter electrode. Any resistive losses, if not negligible, can be determined using the current interruption technique.

The current interruption technique involves passing a current, which is suddenly interrupted by opening a switch, while monitoring the value of V_{WC}-V_{RC}. The resistive losses in the electrolyte decay with a very small time constant and hence the trace exhibits a sudden drop in $V_{WC} - V_{RC}$ (see Figure 4). The remaining potential is the true overpotential of the working electrode. This overpotential then decays with a longer time constant as discharge of the double-layer takes place.

2.3 Charge-transfer mechanism – The nature of the charge-transfer mechanism is important in that under reaction conditions the electroactive species need not be at a constant activity (i.e. there may be activity gradients). It is therefore

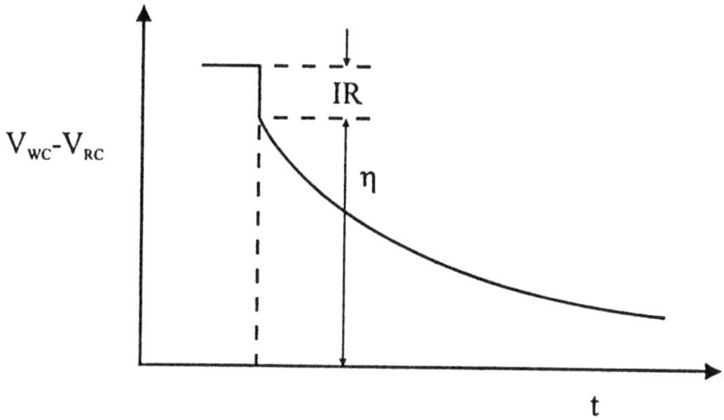

Figure 4 *Schematic showing cell response to current interruption.*

important to know where the charge-transfer reaction is occurring and hence in what location the sensor reflects the activity of the electroactive species.

2.3.1 Metal Electrodes – In the case of metal electrodes, oxygen transfer may take place in the three-phase region, B, at the metal/electrolyte interface, C, or on the electrolyte surface, A, see Figure 5. Because metal electrodes are impermeable to oxygen (with the notable exception of silver), access to reactive sites at the electrode/electrolyte interface, C, involves diffusion along the phase boundary. The dominant mode of oxygen transfer will depend upon the electrode morphology. However, the morphology of the three-phase zirconia/platinum boundary can be influenced by temperature and pressure.[22] Göpel and co-workers report that adsorption and desorption of oxygen from the platinum/ zirconia composite can change the morphology of the interface[23] and result in a

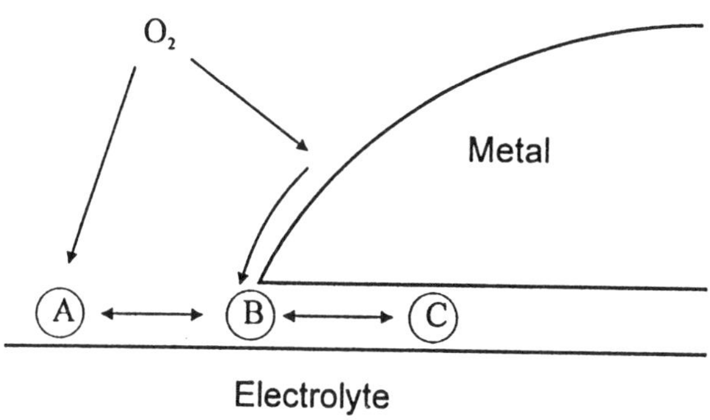

Figure 5 *Schematic showing reaction sites for charge transfer for a metal electrode. A: electrolyte surface. B: three-phase-phase region. C: two-phase electrode/electrolyte interface.*

modification of the kinetic parameters for charge-transfer. Furthermore, interdiffusion is possible at higher temperatures, which in turn will modify the interface properties.

Wang and Nowick have studied the anodic and cathodic reactions of oxygen at platinum electrodes on doped ceria electrolytes.[24-26] At high oxygen partial pressures the current-overpotential relationship indicated that the rate-determining step in the overall process was charge-transfer, which was shown to involve atomic oxygen adsorbed on the platinum surface. At low oxygen partial pressures, limiting currents were observed during cathodic polarisation, while at higher overpotentials, electrolyte reduction appeared to take place, as evidenced by the onset of electronic conduction. In later work,[27] results were consistent with charge-transfer taking place in a two-phase region between the electrode and electrolyte with the rate being determined by the interfacial diffusion of oxygen atoms.

Robertson and Michaels investigated the location of the electrochemical reaction sites for platinum electrodes on zirconia.[28] Experiments were performed with varying partial pressures of oxygen in inert (partial pressures from 10^{-5} to 0.21 atm) from 600 to 800°C. The current-overpotential behaviour and the dynamic electrode behaviour were investigated. No Tafel regions were observed because of the presence of limiting currents. The anodic limiting currents seemed insensitive to the oxygen partial pressure (in disagreement with the work of Mizusaki et al.[29]), whereas, the cathodic limiting current increased as the partial pressure of oxygen was increased. Two different mechanisms for the electrochemical reaction were considered. The first mechanism involved oxygen surface diffusion along the electrolyte/electrode two phase interface as the rate-determining step (as suggested by previous workers[27,30]). This model predicted both anodic and cathodic limiting currents and could be used semiquantitatively to predict the steady-state current-overpotential behaviour. The second mechanism involved reaction at the three-phase boundary. Robertson and Michaels went on to discriminate between the two mechanisms by considering the time dependence of the current in the case of a step change in the overpotential. The experimental current decay was found to agree with the second mechanism, i.e., reaction taking place at the three-phase boundary.

The amount of oxygen adsorbed in the three-phase region has been found to depend linearly on the exchange current density for different catalyst-electrodes under similar conditions.[31,32] This indicates that the electrocatalytic reaction takes place at the three-phase boundary. Vayenas and co-workers pointed out that for less porous electrodes the charge-transfer reaction at the two-phase boundary might become important and that under some conditions oxygen on the electrolyte surface itself might play a role.

2.3.2 Oxide Electrodes – Oxide electrodes are fundamentally different from metal electrodes in that the electrode itself exhibits oxygen-ion conductivity. This means that ionic diffusion through the electrode to the electrode/electrolyte interface may be of importance (see Figure 6).

Mari suggests using oxides as electrode materials[33] in oxygen sensors because

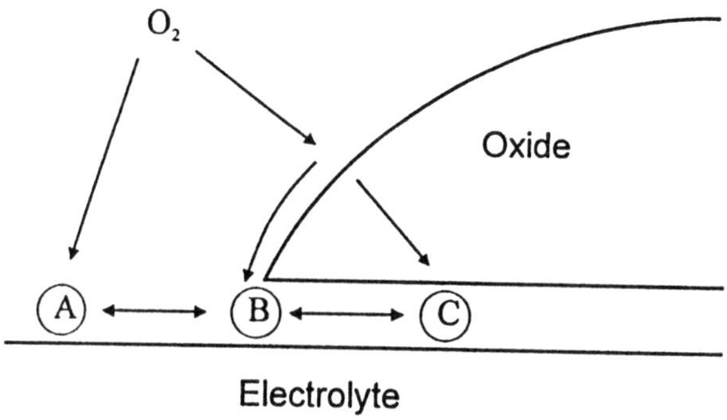

Figure 6 *Schematic showing reaction sites for charge transfer for an oxide electrode. A: electrolyte surface. B: three-phase-phase region. C: two-phase electrode/electrolyte interface.*

oxides equilibrate quickly with gas phase oxygen and mixed ionic and electronic conductivity should improve response times. Mari[33] has summarised work performed on oxide electrodes for oxygen sensor applications and found that, in general, non-stoichiometric oxides work well as reversible electrodes in solid state oxygen sensors. Mixed oxides based on Gd, Zr, Tb, Bi, Ce have been tested as oxide electrode materials between temperatures of 820 and 1020 K in oxygen/inert atmospheres.[34] The form of the current overpotential relationships for the oxide materials gave values of the overall transfer coefficients which indicated that a diffusive process was rate determining, perhaps diffusion in the oxide electrode. It was concluded that both oxygen vacancies and electronic conductivity are necessary for active oxide electrode materials. Petrolekas and Metcalfe[35] investigated a $La(Sr)MnO_3$ electrode using gas phase concentration switching. The long time constants associated with the transient response of the potential indicated that a reaction at the two-phase interface must be important. However, Mizusaki et al.[36,37] have investigated the behaviour of $La(Ca)MnO_3$ cathodes of different morphologies and found no evidence that ionic oxygen transport through the electrode was an important mechanism.

As in the case of metal electrodes, the nature of the charge-transfer reaction for oxide electrodes is a complex function of electrode morphology, overpotential and other conditions.

3 Potentiometric Studies

An early use of oxygen-ion conducting solid electrolytes was in potentiometric devices for the measurement of oxygen partial pressure. In Section 2 it was shown that the e.m.f. of the cell relates the oxygen partial pressures on both sides through the Nernst equation,

$$E_{cell} = \frac{RT}{4F} \ln \frac{P''_{O_2}}{P'_{O_2}} \tag{3.1}$$

This equation relies on the assumption that adsorbed oxygen on the electrode surfaces is in equilibrium with gas phase oxygen. This has lead to a number of applications of this type of cell as a sensor. The principle applications are listed below:[4]

(i) Control of combustion in boilers.
(ii) The monitoring and control of atmospheres in metallurgical heat treatment furnaces.
(iii) Control of the air-fuel ratio for automobile exhaust treatment systems employing a three-way catalyst.
(iv) Control of oxygen partial pressures in laboratory applications.

In the case where one of the electrodes is exposed to a reacting gas mixture this no longer need be true and the e.m.f. of the cell becomes a reflection of the thermodynamic activity of oxygen adsorbed on the electrode in question,*

$$E_{cell} = \frac{RT}{4F} \ln \frac{\alpha_O^2}{P'_{O_2}} \tag{3.2}$$

where α_O is the thermodynamic activity of adsorbed atomic oxygen on the catalyst. In this way, the sensor indicates the surface conditions on the catalyst and not the gas phase conditions. If one is developing a gas sensor then it is important to reduce the effect of reactions on the catalyst surface so as to measure an activity that is close to equilibrium with the gas phase concentration. This is, of course, not the case when the sensor is employed as a catalyst sensor. Under this mode of operation it is important for the sensor to reflect the surface conditions during reaction.

The use of equation (3.2) to study the behaviour of catalysts is known as solid electrolyte potentiometry (SEP). Wagner[38] was the first to put forward the idea of using SEP to study catalysts under working conditions. Vayenas and Saltsburg were the first to apply the technique to the fundamental study of a catalytic reaction for the case of the oxidation of sulfur dioxide.[39] Since then the technique has been widely used, with particular success in the study of periodic and oscillatory phenomena for such reactions as the oxidation of carbon monoxide on platinum, hydrogen on nickel, ethylene on platinum and propylene oxide on silver.

3.1 Metal Catalysts

3.1.1 Total Oxidation

3.1.1.1 Sulfur dioxide oxidation. The first use of SEP to study catalytic processes was made by Vayenas and Saltsburg. Vayenas and Saltsburg studied sulfur dioxide oxidation over platinum, silver and gold electrodes.[39] The oxygen activity

* If mixed potentials occur (see Section 2) then equation (3.2) cannot be applied.

was monitored during SO_2 oxidation at temperatures above 400°C. In general there was no equilibrium between gas phase and adsorbed oxygen.

In the case of platinum three distinct kinetic regimes were observed. At low temperatures the rate of the reaction was limited by the rate of desorption of the product sulfur trioxide; a stable chemisorbed phase of sulfur trioxide was formed, resulting in a poisoning of the reaction rate. At higher temperatures the rate of the reaction was limited by the adsorption of oxygen and poisoning by sulfur trioxide was by the usual site blocking process. Silver and gold films also exhibited a reaction rate limited by the desorption of sulfur trioxide from a stable chemisorbed state. The differences in activities of the three catalysts were explained in terms of differences in their binding strengths with sulfur trioxide. As a result of the work an alternative kinetic model, incorporating this stable sulfur trioxide chemisorbed phase, was proposed, thereby demonstrating the usefulness of SEP work in increasing the understanding of catalytic processes.

3.1.1.2 Hydrogen oxidation. Saranteas and Stoukides[40] studied the oxidation of hydrogen over nickel. Porous polycrystalline nickel films deposited on a stabilized zirconia support were used at temperatures between 240 and 410°C. SEP was used to monitor the chemical potential of oxygen adsorbed on the catalyst. The validity of the Nernst equation under reaction conditions was assumed, i.e. the authors did not consider the possibility of a mixed potential even under very reducing conditions. The authors investigated the rate of reaction using a blank run, i.e. a YSZ tube onto which no nickel had been deposited. It was found that, at the higher temperatures used in the study, the reaction rate was up to 8% of the rate exhibited when the nickel catalyst was present. This rate was explained in terms of a homogeneous reaction.

Reaction rate oscillations were observed with this system. Saranteas and Stoukides studied the oscillations by reducing the catalyst in a hydrogen-nitrogen mixture overnight and then exposing it to reaction mixtures (this form of pretreatment was not used in the rest of the study). Oscillations did not appear until after 30-40 minutes of operation under reaction conditions. An interesting result is that both oscillations in the outlet partial pressure of oxygen and the e.m.f. were noted. However, the two forms of oscillation did not always occur simultaneously; in some cases the outlet oxygen partial pressure oscillated while the e.m.f. remained constant; the opposite phenomenon was also observed.

Saranteas and Stoukides explained their results in terms of two regions, one in which the nickel surface is primarily oxidised and one in which the nickel surface is primarily reduced, although under reaction conditions nickel and its oxide can coexist. High reaction rates were obtained on the clean nickel surface while very low chemical potentials of oxygen were observed. As the partial pressure of oxygen was increased the oxide began to form, the chemical potential of oxygen increasing and the rate dropping.

Two models were presented for the steady state oxidation of hydrogen over nickel. Both include a step for the formation of nickel oxide, one from adsorbed oxygen and the other from gaseous oxygen. The equilibrium constants for these two processes were calculated so as to discriminate between the models, the conclusion being that oxygen must first adsorb before forming the oxide. However, the analysis relies upon the assumption that the adsorbed oxygen is in equilibrium with the oxide, which need not be the case under reaction conditions.

E.m.f. oscillations during the oxidation of hydrogen over nickel have been studied further by Stoukides and co-workers.[41,42] Oscillations occurring in the temperature range 510 to 618 K were also explained in terms of the formation and destruction of surface nickel oxide.[42]

3.1.1.3 Carbon monoxide oxidation. Hetrick and Logothetis were among the first to study the oxidation of CO over Pt using SEP.[43] Hetrick and Logothetis considered the possibility that, in the presence of carbon monoxide, a mixed potential may occur. At high temperatures (greater than 500°C) the reactant gases equilibrate and the e.m.f. data could be explained by use of the Nernst equation. A sharp transition in the e.m.f. signal was observed at the stoichiometric point (this property has resulted in the use of such sensors in monitoring and controlling air/fuel ratios in automobiles). At lower temperatures equilibration was not observed. Self sustained oscillations were observed at these lower temperatures with gas phase concentrations close to the stoichiometric point. Simultaneous measurements of the rate of carbon dioxide production and the e.m.f. indicated that the electrical oscillations occurred in phase with oscillations in the coverages of adsorbed carbon monoxide and oxygen. The authors also noted that they could modify the nature of the oscillations by passing a current through the cell.

Haaland[44] carried out work on the use of oxygen-ion conducting solid electrolyte cells as oxygen sensors using what was described as non-catalytic electrodes. These electrodes were made of platinum, gold or silver and were suitably modified with lead and sulfur to try to minimise any catalytic activity. Haaland carried out a study of the reaction mechanism for CO oxidation over a partially poisoned lead/platinum electrode. It was claimed that because the reaction rate between CO and oxygen is reduced by the poison, the e.m.f. of the cell becomes a measure of the oxygen partial pressure in the gas phase near to the electrode surface. In this way the cell can be used as a probe to study reaction mechanisms. Based on this work Haaland proposes the rather unlikely mechanism whereby gas phase CO attacks oxygen adsorbed on the platinum surface via an Eley-Rideal mechanism.

Okamoto and co-workers have intensively studied the oxidation of CO on platinum[45-52] using SEP. In their early work,[45] Okamoto et al. investigated carbon monoxide oxidation using a potentiometric device incorporating a YSZ electrolyte at temperatures around 350°C. Under oxygen rich conditions they observed an anomalously high e.m.f. which they explained by the presence of a mixed potential involving carbon monoxide. The study was rather limited in that no kinetic data were recorded, only e.m.f. data. The main focus of the work, rather than to properly explain the anomalous behaviour in the presence of CO, was to develop a new type of CO sensor for detecting low concentrations of CO in air without the use for a separate reference gas supply. Okamoto and co-workers deposited platinum electrodes on both sides of a YSZ membrane and onto one of these platinum electrodes they deposited a porous CO oxidation catalyst consisting of platinum supported on alumina. The idea was that the electrode covered with the porous CO oxidation catalyst would not be exposed to CO, as the CO would be consumed by the reaction in the catalyst pores. Hence this electrode would serve as a suitable reference electrode. Indeed the sensor did appear to work as a CO detector for concentrations at around the 100 ppm level.

In later work, Okamoto and co-workers attempted to explain the e.m.f. in the presence of CO and oxygen mixtures. This study[49] included the use of an infrared cell into which a pellet of platinum supported on YSZ was placed. A differentially operated fixed bed reactor was used to determine the reaction kinetics for CO oxidation over platinum supported on YSZ. The temperature of the electrode surface was measured by attaching a thermocouple. The fact that the e.m.f. data, the infrared data and the kinetic data were all measured in three different systems did not help analysis. Results for e.m.f. and catalyst temperature rise (indicative of the reaction rate) were plotted against gas phase composition for the platinum electrode. As with other work, the kinetic results fell into three regimes. A fuel rich regime, in which high e.m.f.s and low reaction rates were observed, an oxygen rich regime, in which low e.m.f.s and high reaction rates were observed, and an intermediate regime in which e.m.f. oscillations were observed. In the case of Okamoto and co-workers no reaction rate oscillations were observed in the intermediate regime, perhaps because of the long time constant associated with the thermal response of the system. The kinetics in regimes I and III (oxygen rich and CO rich) were found to be similar to those observed by previous workers.[53] Okamoto and co-workers concluded that, based on the e.m.f. data, the potential was a mixed one, the following charge-transfer reactions being dominant:

$$* + O_O \Leftrightarrow O^* + V_O^{\cdot\cdot} + 2e' \qquad (3.3)$$

$$CO^* + O_O \Leftrightarrow CO_2 + * + V_O^{\cdot\cdot} + 2e' \qquad (3.4)$$

However, the analysis of the data was carried out in such a way as to cast doubt on the validity of these conclusions (Vayenas[54]). Okamoto, Kawamura and Kudo[48] went on to use the e.m.f. interpretation from the above work[49] to further investigate the mechanism of CO oxidation over platinum by using the cell as a probe of the surface coverage of carbon monoxide.

Earlier work was complemented with exchange current density measurements.*[51,52] E.m.f. and current density data were inconsistent with only one charge-transfer reaction involving oxygen. Again, two reversible charge-transfer reactions were considered, (3.3) and (3.4). The anodic branch of (3.3) was neglected as was the cathodic branch of (3.4) based on the approximate values of the e.m.f. of the cell compared to the standard reversible potentials (however,

* Although it is necessary to polarise a cell to obtain the exchange current density, the exchange current density refers to an unpolarised electrode and therefore can be considered as potentiometric information. Furthermore, if exchange current density is to be measured this is more easily achieved if the charge-transfer reaction is slow. In such a case the resistance measured will be due to charge transfer alone and analysis is straightforward.

such potentials are of no particular significance under reaction conditions, as the surface of the electrode is not necessarily close to equilibrium with the gas phase). The work must be treated cautiously because of the possibility of the anodic branch of (3.3) occurring under the conditions employed. However, the principle of using both e.m.f. and current density measurements simultaneously to investigate the behaviour of catalysts is a valid one and has since been used to investigate the same system.[15]

Metcalfe and Sundaresan[15] went on to confirm that e.m.f.s obtained in the case of CO oxidation over platinum cannot be explained if reaction (3.3) is the only electrochemical reaction to occur. Even under oxygen rich conditions, carbon monoxide was involved in the electrochemical reactions, the dominant electrochemical reactions being,[15]

$$* + O_O \Leftrightarrow O* + V_O^{\cdot\cdot} + 2e' \tag{3.5}$$

$$CO* + O_O \rightarrow CO_2* + V_O^{\cdot\cdot} + 2e' \tag{3.6}$$

Although only one cathodic reaction was occurring, two anodic reactions were present.

Under CO rich conditions the behaviour was more complicated due to the involvement of the electrolyte surface through the cathodic reaction,

$$O*_{YSZ} + V_O^{\cdot\cdot} + 2e' \rightarrow *_{YSZ} + O_O \tag{3.7}$$

where $*_{YSZ}$ represents a site on the electrolyte surface. Metcalfe and Sundaresan were able to show that this was the only cathodic reaction taking place under CO rich conditions. The only anodic reaction was reaction (3.6), as in the oxygen rich regime. If reactions (3.6) and (3.7) are added together the net reaction is,

$$CO* + O*_{YSZ} \rightarrow CO_2* + *_{YSZ} \tag{3.8}$$

i.e. the electrochemical reactions allow carbon monoxide adsorbed on the platinum to react with oxygen adsorbed on the surface of the electrolyte.† This process was shown to be of a sufficient magnitude that it may modify the behaviour of a catalyst if small platinum crystallites (10-50 Å) were present.

Lintz and co-workers have investigated the effect of a bismuth additive in metal electrodes for carbon monoxide oxidation[56] and have also used SEP to study carbon monoxide oxidation.[57]

Potentiometric techniques have been used to study autonomous reaction rate oscillations over catalysts and carbon monoxide oxidation on platinum has received a considerable amount of attention[43,48,58] Possible explanations for reaction rate oscillations over platinum for carbon monoxide oxidation include, (i) strong dependence of activation energy or heat of adsorption on coverage, (ii) surface temperature oscillations, (iii) shift between multiple steady states due to adsorption or desorption of inert species, (iv) periodic oxidation or reduction of the surface. The work of Sales, Turner and Maple has indicated that the most

† This reaction has also recently been shown to proceed directly, i.e., without the need for charge-transfer, in the three-phase region.[55]

probable explanation is due to the periodic oxidation and reduction of the surface.[59,60]

Yentekakis and co-workers studied the oxidation of CO over platinum, including the associated autonomous oscillations, using SEP.[58] Surface titrations of both adsorbed CO and adsorbed oxygen were used to measure the electrode area. Evidence from these experiments indicated that oxygen was undergoing multilayer adsorption at higher temperatures (above 500°C), i.e. platinum oxide formation was occurring. Under CO rich conditions, the involvement of CO in the electrode charge-transfer reactions leading to a mixed potential could not be ruled out, however, Vayenas states that the e.m.f. would still provide a qualitative measure of the oxygen activity under these CO rich conditions (this could be disputed as there is evidence that the dominant cathodic reaction in this region may not actually involve oxygen adsorbed on the platinum surface at all[15]). However, the bifurcation between oscillatory and non-oscillatory states occurs near to the dissociation pressure of platinum oxide. Vayenas indicates that this dissociation pressure for platinum oxide is well within the fuel-lean region of operation and that hence the interpretation of the e.m.f. is unambiguous as only one charge-transfer process will proceed.

The catalyst[58] was left in either air or CO overnight before carrying out the kinetic experiments. At temperatures above 400°C there was little effect due to the pretreatment of the catalyst, the surface responding quickly to changes in gas phase conditions. Below this temperature the surface had a long memory effect exhibited by the way in which the reaction rate depended upon the pretreatment. At these temperatures reaction rate oscillations were also observed (see Figure 7[58]). On a pre-oxidised surface e.m.f. spikes corresponded to peaks in the oxygen activity and coincided with peaks in the reaction rates. The opposite was true of the form of the oscillations on the prereduced catalyst. The effect of gas phase conditions was investigated by keeping the partial pressure of oxygen constant and varying the partial pressure of CO. On the pre-oxidised surface a low partial pressure of CO resulted in a high baseline oxygen activity with a low amplitude for the oscillations. As the partial pressure of CO was increased, the baseline oxygen activity dropped and the high oxygen activity excursions occurred with a higher frequency. Similarly when the pre-reduced surface was exposed to a low partial pressure of CO low amplitude oscillations were observed with a high baseline oxygen activity close to that of the pre-oxidised case. Unlike the pre-oxidised case, when the partial pressure of oxygen was increased, the baseline oxygen activity remained approximately constant; the amplitude of oxygen activity excursions increased as did their frequency. Vayenas and co-workers studied the synchronization of the oscillations by investigating the behaviour of two electrodes. The two catalyst electrodes behaved in a similar manner. This was explained by changes in the gas phase composition rapidly affecting conditions on the catalyst surface and hence the two electrodes communicated via the gas phase.

Results were modelled on the basis of CO adsorption/desorption equilibrium with oxygen adsorption as the rate determining step and an irreversible Langmuir-Hinshelwood surface reaction. It was assumed that adsorption and

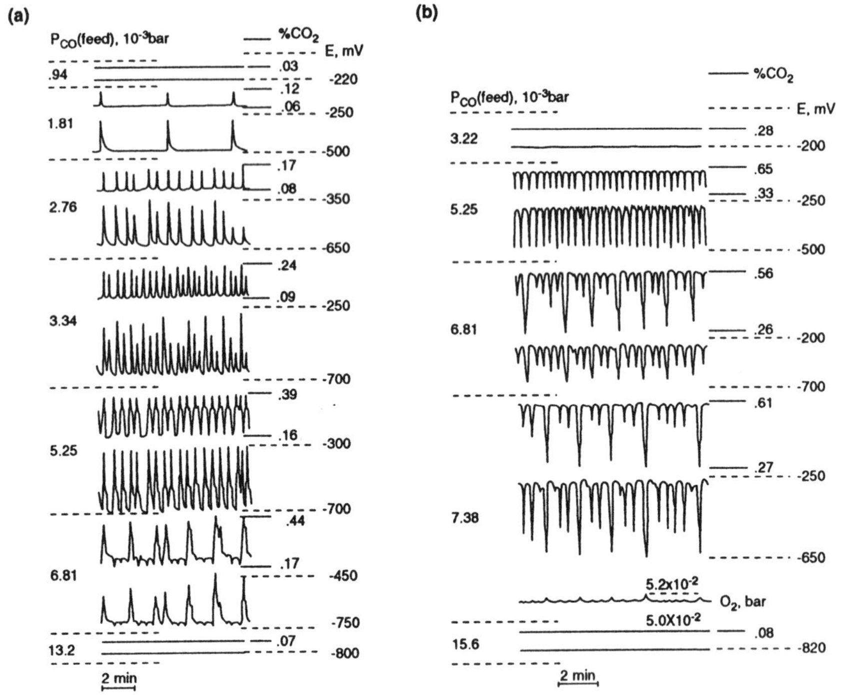

Figure 7 *Rate and e.m.f. oscillations for CO oxidation over platinum at 337 °C and inlet* P_{O2} = *5.43* × 10^{-2} bar (from 58). (a) *Oscillations obtained on a pre-oxidised surface*; (b) *oscillations obtained on a pre-reduced surface.* (Reproduced from *J. Catal.*, 1988, **11**, 152, by permission from Academic Press Inc.)

reaction did not take place on oxidised surface sites. The kinetics of the formation and destruction of the oxide were not included and hence the model could only be used to fit the steady-state results. To fit the model to both pre-oxidised and pre-reduced results, two different adsorption/desorption equilibrium constants had to be used. Analysis of the temperature dependence of the model indicated that the entropy of dissociative oxygen adsorption was positive. This inconsistency was taken to indicate that there must be a mixed potential in the fuel-rich regime.

3.1.1.4 Methane oxidation. In early work[61] Sandler investigated the behaviour of a calcia-stabilized zirconia (CSZ) cell with platinum and silver electrodes in the presence of methane-oxygen mixtures. The results indicated that the cell could be used as a methane detector device. Haaland[44] also investigated the behaviour of platinum, silver and gold electrodes in the presence of methane-oxygen mixtures under open-circuit conditions. Platinum was found to be the most active electrode for methane oxidation and SEP was used to investigate the reaction mechanism.

Seimanides and Stoukides have carried out some more recent work on the oxidation of methane.[62,63] In the first of these papers,[62] methane oxidation over polycrystalline silver was studied (no partial oxidation products were observed in this study). In general, equilibrium did not exist between adsorbed oxygen and gas phase oxygen during methane oxidation between 450 and 600°C. However, differences in α_O and $P_{O2}^{1/2}$ were much less for methane oxidation than in the cases of ethylene and propylene oxidation (this is consistent with the relatively lower rates in the case of methane oxidation). A model was presented which accounted both for the kinetic behaviour and the dependence of the oxygen activities on the gas phase conditions. The reaction was suggested to proceed via an Eley-Rideal mechanism with adsorbed oxygen reacting with gas phase methane. Similar results were obtained with a palladium catalyst.[63] More recent work on methane oxidation over both silver and Ni/ZrO_2 catalysts supported on YSZ at 600-800°C,[64] has indicated that mixed potentials are important and should be considered when interpreting SEP results.

3.1.1.5 Ethylene and ethylene oxide oxidation. Oscillations during the oxidation of ethylene over platinum[65,66] have been studied. Vayenas and co-workers studied the complete oxidation of ethylene over platinum (no conversion to partial oxidation products was observed, and the zirconia tube was shown to be inactive for the reaction) in the temperature range of 200-400°C at atmospheric pressure.[65] By varying feed flow rates the authors demonstrated that there were no mass transfer resistance effects. Internal mass transfer in the pores of the electrode was assumed to be negligible as previous work had shown that the chemical potential of oxygen at the three-phase interface did not depend upon the thickness of the electrode.[39] SEP was used to monitor the oxygen activity on the platinum catalyst. Oscillations in the reaction rate occurred in a particular range of partial pressures; this was explained in terms of the stability of platinum oxide. The chemical potential of the adsorbed oxygen was found to be dependent on both the partial pressure of oxygen and the partial pressure of ethylene in the following way,

$$\alpha_O = K \frac{P_{O_2}}{P_{C_2H_4}} \tag{3.0}$$

where K is a function of temperature alone.

When the partial pressure of ethylene was low the rate of reaction was first order in ethylene and zero order in oxygen. At higher partial pressures of ethylene (ethylene to oxygen ratios of greater than 1:3), the rate was first order in oxygen and zero order in ethylene and self-sustained reaction rate oscillations were observed. From e.m.f. measurements, the oscillations were observed to occur in a fixed region of oxygen chemical potentials, the frequency and amplitude of the oscillations being dependent upon the oxygen chemical potential.

The kinetic model derived to fit the experimental observations depended upon gaseous ethylene reacting with adsorbed atomic oxygen forming a highly reactive intermediate which then undergoes decomposition to form the total oxidation products. The rate-determining step involved the adsorption of molecular oxygen which then rapidly equilibrates with surface atomic oxygen. Such a model agreed with both the kinetic data and the data on the chemical potential of adsorbed oxygen. The existence of self-sustained reaction rate and e.m.f. oscillations was explained in terms of the formation and destruction of surface platinum oxide. The lower limit of oxygen chemical potential for the observance of oscillations corresponded to the minimum surface oxygen activity required for the existence of platinum oxide.

Stoukides and Vayenas[67] have also studied the oxidation of ethylene oxide in the region of 250-400°C. Stoukides and Vayenas proposed a model whereby ethylene oxide dimerizes before undergoing reaction with adsorbed oxygen to form total oxidation products. Oxygen adsorbs in a precursor state on an S2 surface site. Ethylene oxide can only adsorb on an S1 surface site. The rate determining step for the reaction is the migration of adsorbed atomic oxygen from an S2 site to the S1 site. Unfortunately, discussion of the final model was somewhat limited.

3.1.2 Partial Oxidation

3.1.2.1 Oxidative coupling of methane. SEP and electrochemical oxygen pumping (EOP) were used in the investigation of the partial oxidation of methane over a platinum electrode[68] between 600 and 900°C at atmospheric pressure. A tubular configuration was used, reaction rates being determined by the difference in inlet and outlet concentrations (although there must be some question as to how representative of the conditions within the reactor these concentrations are when the reactor is operated integrally). In the outlet stream, carbon dioxide, carbon monoxide, water, hydrogen and both ethane and ethylene were observed. Kinetic measurements at 600, 700, 800 and 900°C were made in conjunction with SEP measurements. At both 600 and 700°C formation rates of C_2s were negligible. The SEP results were used to determine the oxygen coverage on the platinum catalyst as in previous work.[69] From the reaction rate dependence on the methane partial pressure, both the homogeneous and the heterogeneous reactions were concluded to be important. The activation energies for both the surface and gas-phase reactions were found to be the same and it was suggested that this may be due to a common rate-determining step, this being hydrogen abstraction from surface methane to form a methyl radical.

3.1.2.2 Ethylene epoxidation. Stoukides and Vayenas studied the epoxidation of ethylene over silver between 250 and 450°C at atmospheric pressure.[70] Three reactions occurred,

$$C_2H_4 + \tfrac{1}{2}O_2 \rightarrow C_2H_4O \qquad (3.10)$$

$$C_2H_4 + 3O_2 \rightarrow 2CO_2 + 2H_2O \qquad (3.11)$$

$$C_2H_4O + \tfrac{3}{2}O_2 \rightarrow 2CO_2 + 2H_2O \qquad (3.12)$$

The presence of carbon dioxide was found to inhibit the rate of reaction (3.10) but did not appreciably affect the rate of reaction (3.11). Oxygen activity data could be accounted for if two different types of adsorbed oxygen were present. Carbon dioxide competed for sites with only one of these forms of adsorbed oxygen, i.e., the form responsible for ethylene oxide formation. It is probable that adsorbed atomic oxygen was responsible for the formation of carbon dioxide and water with molecular oxygen being responsible for the formation of ethylene oxide. Stoukides and Vayenas suggested that it may be possible for Ag_2CO_3 to be formed by carbon dioxide adsorbing on the atomic oxygen left behind after the molecular oxygen has reacted to form ethylene oxide. This Ag_2CO_3 could then block sites for molecular oxygen adsorption.

3.1.2.3 Propene epoxidation. Stoukides and Vayenas have studied the epoxidation of propene over silver catalysts.[71-73] A Langmuir-Hishelwood type model was used to explain the results of work performed between 290 and 400°C.[71] As with the work on ethylene oxidation, two types of oxygen were proposed to be involved, molecular and atomic oxygen responsible for partial and total oxidation respectively.

Rate oscillations during the oxidation of propylene oxide, the desired product of the above reaction, on polycrystalline silver were studied using an gas phase CSTR operating at temperatures between 250 and 400°C.[74] Oscillations in the reaction rate were found to be synchronous with oscillations in the oxygen activity, with high rates of carbon dioxide formation corresponding to high surface oxygen chemical potentials. A model was presented which explained the steady-state behaviour of the reaction, however, there was no correlation between the surface oxygen activity and the limits of the oscillatory behaviour. The oscillations were explained by propylene forming relatively stable intermediates, evidence for the isomerization to acrolein and polymer formation[75] being cited.

3.1.3 Other Reactions

3.1.3.1 Nitric oxide reduction with carbon monoxide. This reaction, which is of great importance in automobile exhaust catalysis, has received very little attention. The only work using SEP as an in-situ probe to study the catalytic reaction was performed by Häfele and Lintz[76,77] who worked with a platinum catalyst. High surface oxygen activities were measured at low carbon monoxide partial pressures and the reaction rate had a first order dependence on the partial pressure of CO. At higher partial pressures of CO, there was a second kinetic regime characterized by low surface oxygen activities and a negative order dependence of reaction rate on the CO partial pressure. At high partial pressures of NO (0.08 bar) the rate of reaction was decreased by increasing levels of NO. This was explained by the presence of undissociated NO on the catalyst resulting in less atomically adsorbed oxygen as demonstrated by lower oxygen activities and lower reaction rates.

3.1.3.2 Nitric oxide reduction with ammonia. Lintz and co-workers[78,79] have studied the platinum catalysed reduction of nitric oxide by ammonia in the region of 300-400°C. Three reactions were shown to proceed; the direct reduction of nitric oxide to nitrogen,

$$6NO + 4NH_3 \rightarrow 5N_2 + 6H_2O \qquad (3.13)$$

and the reduction of nitric oxide to nitrous oxide followed, potentially, by the further reduction of the nitrous oxide:

$$8NO + 2NH_3 \rightarrow 5N_2O + 3H_2O \qquad (3.14)$$

$$5N_2O + 10/3\ NH_3 \rightarrow 20/3\ N_2 + 5H_2O \qquad (3.15)$$

It was found that when the partial pressure of nitric oxide was more than 1.5 times greater than the partial pressure of ammonia, the platinum surface remained in a relatively oxidised state and the selectivity to nitrogen was poor. When the partial pressure of nitric oxide was less than 1.5 times the partial pressure of ammonia, the platinum surface remained in a relatively reduced state and the selectivity to nitrogen approached unity.

3.2 Oxide Catalysts – Under appropriate conditions a bulk metal will form an oxide overlayer and this layer will be responsible for governing the catalytic behaviour. In a similar manner a bulk oxide can undergo reduction with a formation of a metallic surface layer. Such behaviour can be responsible for rate oscillations or hysteresis in reaction rate and is important when considering the catalytic behaviour of bulk oxides.

SEP data during ethylene oxidation over platinum[80] indicate that oscillations in oxygen activity occur in the vicinity of a critical value of the oxygen activity. This has been explained by the formation and destruction of a surface oxide,

$$Pt(s) + \frac{1}{2}O_2(g) \Leftrightarrow PtO(s) \qquad (3.16)$$

The heat and entropy associated with the above reaction implied by the temperature dependence of the critical oxygen activity were -46 kcal mol O_2^{-1} (-192 kJ mol O_2^{-1}) and -45 kcal K^{-1} mol O_2^{-1} (-188 kJ K^{-1} mol O_2^{-1}) in close agreement with the results of Berry.[81] Oxygen adsorption and oxide formation on platinum have also been studied using exchange currents in a solid electrolyte cell.[82]

Figure 8 shows how hysteresis can occur under reaction conditions in the region of a metal/metal oxide transition. Surface oxygen activity is shown as a function of gas phase oxygen pressure for both the metal and metal oxide phases. If equilibrium existed between the surface and gas phase then the square of the oxygen activity would equal the oxygen partial pressure. However, under reaction conditions the surface oxygen activity must be lower than its equilibrium value. The actual relationship between the oxygen activity and partial pressure is kinetically determined for the two phases. The square of the critical oxygen activity, $(\alpha^*_O)^2$ corresponds to the decomposition pressure of the oxide. It can be seen that the history of the catalyst will influence the rate of reaction at oxygen partial pressures between $P_{O2,1}$ and $P_{O2,2}$ due to the presence of a hysteresis loop

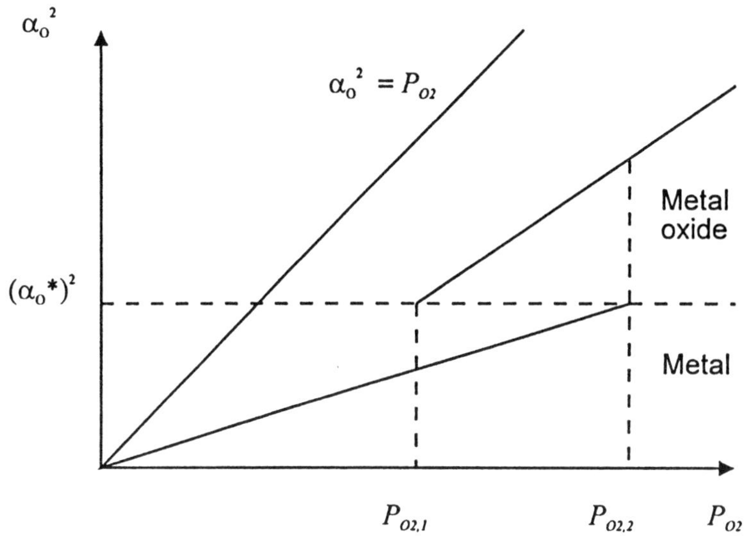

Figure 8 *Schematic representation of the relationship between oxygen activity (α_O) and oxygen partial pressure* (P_{O2}) *for a metal/metal oxide catalyst.*

as both phases are stable within this oxygen partial pressure range. Consider the metal exposed to an oxygen partial pressure below $P_{O2,1}$. As the oxygen partial pressure is increased the metal remains the stable phase until a pressure of $P_{O2,2}$ is reached when the oxide is formed. If we now reduce the oxygen partial pressure the oxide will remain the stable phase until we reach a partial pressure of $P_{O2,1}$ when once more the metal is formed. It is also possible in metal/metal oxide systems for there to be a range of oxygen partial pressures in which no one phase is stable and a diagram similar to Figure 8 can be constructed. The two phases then coexist as a dynamic patchwork on the catalyst surface and oscillations in reaction rate are observed.

3.2.1 Total Oxidation

3.2.1.1 Hydrogen oxidation. Balian et al.[83] have studied hydrogen oxidation over copper and copper oxide catalysts using potentiometric techniques. Experiments were performed in the temperature range of 250-365 °C at atmospheric pressure. The YSZ cell was used to monitor the stable phase of the solid catalyst. Under most conditions employed, CuO was the stable phase of the catalyst. Under conditions for which CuO was the stable phase the e.m.f. of the cell was between −40 and −180 mV and therefore it was assumed that a mixed potential involving hydrogen in a charge-transfer reaction was unimportant. The reaction on the CuO surface was found to follow a Langmuir-Hinshelwood mechanism with atomic oxygen adsorption being much stronger than that of hydrogen. The reaction of adsorbed hydrogen with lattice oxygen did not appear to be significant in this regime.* Under conditions in which Cu_2O and Cu were the stable catalyst phases, water production was due both to reaction between adsorbed species and oxide reduction by hydrogen. The kinetics were slowest on the CuO and fastest on the Cu metal. Due to the combination of temperatures and partial pressures used, it was easier to study the CuO/Cu_2O phase transition rather than Cu_2O/Cu transition. However, oscillatory phenomena were not observed under any conditions; instead stable steady states were always obtained. In contrast to the work of Greger et al.[84,85] on propene oxidation over copper oxide no hysteresis associated with the phase transition was mentioned.

3.2.1.2 Carbon monoxide oxidation. SEP was used to investigate a $La(Sr)MnO_{3-\delta}$ catalyst during carbon monoxide oxidation at 500 and 550°C.[35] This material was chosen because of its phase stability throughout the conditions employed. Changes in composition and temperature do not result in phase changes, rather they modify the value of δ.

At 500°C the reaction rate over the platinum electrode-catalyst appeared to be independent of the e.m.f. of the cell. At 550°C two reaction rate branches were observed, depending on whether the catalyst had been pretreated in oxidising or reducing conditions (see Figure 9). The e.m.f. of the cell also exhibited two branches dependent upon pretreatment (see Figure 9), in a similar manner to other SEP work on oxide catalysts.[86,87] It was suggested that the catalyst state (i.e., catalyst oxygen content or δ) was a function of the catalyst history. Different catalyst states corresponded with different catalyst reactivities and the e.m.f. of the cell reflected the catalyst state.

The authors discussed in detail where the dominant charge-transfer reaction was taking place. It was possible that charge-transfer was important throughout the two-phase interface as well as within the three-phase region. Long time constants associated with e.m.f. transients indicated that there was a significant contribution from the two phase region. From the transient work it was concluded that the e.m.f. did not reflect surface conditions.

The kinetic results were explained in term of an 'ionic redox model' at 550°C. In this model oxygen reacts with vacancies at sites responsible for the reoxidation of the catalyst. Oxygen is then transported by lattice diffusion to sites responsible for catalyst reduction where the oxygen reacts with adsorbed carbon monoxide to form carbon dioxide. The e.m.f. at steady state is a reflection of the thermodynamic activity of the ionic oxygen species involved in the mechanism by migration between the two sites. The overall rate of reaction can then be expressed in terms of catalyst oxidation or reduction depending upon whether the ionic oxygen is in equilibrium with oxygen at the sites responsible for catalyst oxidation or reduction respectively.†

* The formation of OH groups was not considered in the mechanism although the workers did indicate that an equally valid mechanism involving OH surface species could be formulated that would explain the experimental results.

† If the ionic oxygen were in equilibrium with both oxygen at the sites responsible for catalyst oxidation and reduction, then the model would reduce to a Mars-van Krevelen model and hysteresis and multiple steady states could not be explained.

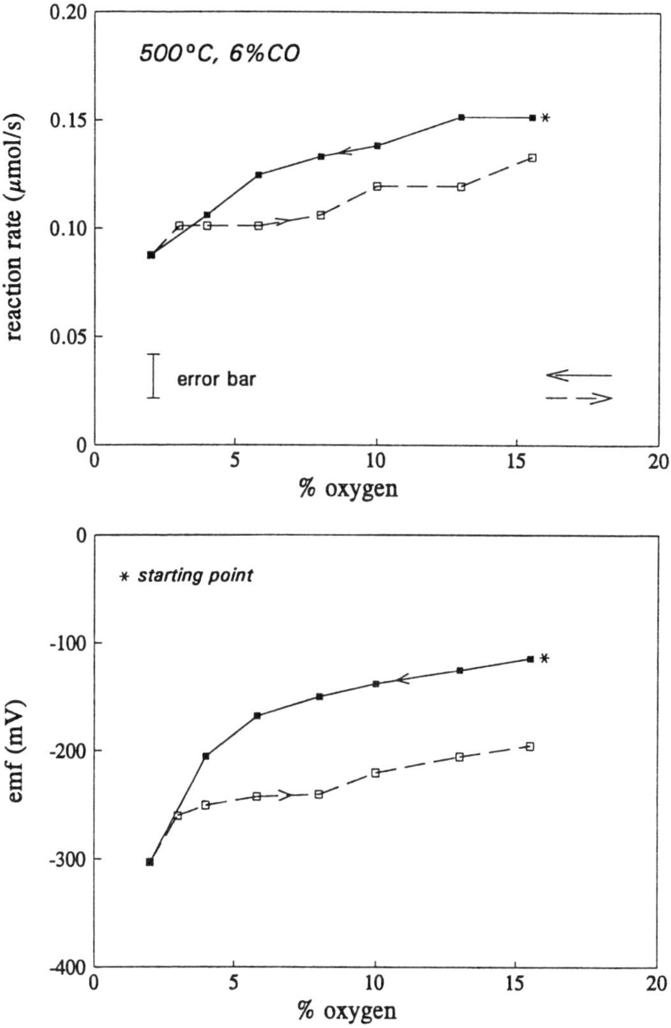

Figure 9 *Steady-state rate and e.m.f. measurements as a function of oxygen in feed during CO oxidation at 500°C and 6%CO over a La(Sr)MnO_3 catalyst (from 35). Closed (open) symbols signify the measurements at decreasing (increasing) percentages of oxygen.*
(Reproduced from *J. Catal.*, 1995, **152**, 147, by permission from Academic Press Inc.)

At 500°C, the reaction is confined to the surface of the catalyst and can be described with Mars-van Krevelen kinetics. The rate is insensitive to the state of the catalyst bulk and hence there was no relationship between e.m.f. and rate.

3.2.2 Partial Oxidation

3.2.2.1 Propene to acrolein. Hildenbrand and Lintz[87,88] have used solid electrolyte potentiometry to study the effect of the phase composition of a copper oxide catalyst on the selectivity and yield of acrolein during the partial oxidation of propene in the temperature range of 420-510°C. Potentiometric techniques were used to determine the catalyst oxygen activity, and hence the stable copper phase, under working conditions. Hildenbrand and Lintz used kinetic measurements to confirm that the thermodynamically stable phase had been formed (it is known that propene is totally oxidised over CuO but partially oxidised over Cu_2O).

The results of Hildenbrand and Lintz showed good quantitative agreement with previous kinetic work of Riekert and Greger.[84,85,89] Reaction rate measurements were indicative of which copper phase was present; this phase corresponding to the thermodynamically favoured phase. Furthermore, hysteresis observed in the reaction rate data was also observed in the oxygen activity measurements as in other SEP work on oxides.[35,86]

3.2.2.2 Maleic anhydride production. The oxidation of benzene to maleic anhydride over a vanadium pentoxide electrode has been studied by Pizzini et al.[90,91] Unfortunately, the quantities of benzene and maleic anhydride were not determined experimentally. Breckner et al.[86] have studied the partial oxidation of butene to produce maleic anhydride over a vanadium phosphate catalyst. Reaction rate and oxygen activity were monitored in order to correlate catalyst selectivity with oxygen activity. The selectivity of the catalyst was found to increase as the oxygen activity of the catalyst decreased. Both the catalyst reactivity and oxygen activity were found to be dependent upon prior treatments.

However, their work was complicated by the fact that they used a porous platinum current collector underneath the oxide. If the gaseous atmosphere were able to come in contact with this porous platinum layer then, under reaction conditions, the platinum could exhibit a different oxygen activity from the vanadium phosphate making the e.m.f. of the cell extremely difficult to interpret.

3.2.2.3 Synthesis gas production. Alqahtany et al.[92] have studied synthesis gas production from methane over an iron/iron oxide electrode-catalyst. Although the study was essentially devoted to fuel cell operation, for purposes of comparison some potentiometric work was performed at 950°C. It was found that under reaction conditions Fe, FeO or Fe_3O_4 could be the stable catalyst phase. Hysteresis in the rates of methane conversion were observed with much greater rates over a pre-reduced surface than over a pre-oxidised surface possibly due to the formation of an oxide.

4 Amperometric Studies

In this section the use of amperometric techniques for the in-situ study of catalysts using solid state electrochemical cells is discussed. This requires that the potential of the cell is disturbed from its equilibrium value and a current passed. However, there is evidence that for a number of solid electrolyte cell systems the change in electrode potential results in a change in the electrode-catalyst work function.[5] This effect is known as the non-faradaic electrochemical modification of catalytic activity (NEMCA). In a similar way it appears that the electrode potential can be used as a monitor of the catalyst work function. Much of the work on the closed-circuit behaviour of solid electrolyte electrochemical cells has been concerned with modifying the behaviour of the catalyst (reference 5 is an excellent review of this area). However, it is not the intention of this review to cover catalyst modification, rather the intention is to address information derived from closed-circuit work relevant to an unmodified catalyst surface.

4.1 Steady-state Current Overpotential Behaviour – For a simple single charge-transfer process equation (2.28) describes the closed-circuit behaviour. At low overpotentials, the current and overpotential are linearly related and the exchange current density can be evaluated from the gradient (see equation (2.25)). As the overpotential is increased the current will rise until it eventually reaches a plateau corresponding with a limiting current. At this point a process other than the charge-transfer process will determine the rate, and overpotential will no longer influence the kinetics. As an example, if we consider the cathodic reduction of oxygen, the three-phase region may, under limiting current conditions, be devoid of surface oxygen. The current is determined by the rate of gas phase diffusion and adsorption in the region or the rate of surface diffusion to the region and the limiting current will depend upon the partial pressure of oxygen. The overall kinetic constant will either equal the oxygen adsorption rate constant (if adsorption is rate determining) or the mass transfer coefficient (if gas phase transport is rate determining). If limiting currents are used properly they could provide useful information regarding the behaviour of a catalyst, however, their use requires a high level of existing knowledge regarding the mechanistic and kinetic behviour of the catalyst. It is probable that such requirements have prevented any significant use of limiting currents for catalyst sensing.

4.2 Cyclic Voltammetry – Cyclic voltammetry is a very good technique for initial electrochemical studies and for work on more complex electrochemical systems. It has been widely used in aqueous electrochemistry. The technique involves the ramping of the cell voltage using a triangular wave generator. Sweep rates of the order of 1 mV s^{-1} to 1 V s^{-1} tend to be used. The current is recorded against the cell voltage, this plot being known as a cyclic voltammogram. The form of the cyclic voltammogram is characteristic of the electrode processes occurring. Consider the reversible charge transfer reaction (reversibility, in this context, means that the charge-transfer processes are fast enough to remain in equilibrium and therefore the Nernst equation will still govern the surface coverages),

$$* + O_O \Leftrightarrow O^* + V_O^{\cdot\cdot} + 2e' \qquad (4.1)$$

Current may be recorded against applied potential for varying sweep rates.[93] At low sweep rates steady state current-overpotential behaviour is seen. As sweep rate is increased a peak of increasing height is observed. As the potential of the electrode is made more negative the concentration of oxygen on the electrode surface is reduced and a concentration gradient develops between oxygen in the charge-transfer region and oxygen on the rest of the electrode, hence, the passage of current increases. As the coverage of oxygen in the charge-transfer region approaches zero, the current can no longer increase and therefore reaches a maximum. At more negative potentials, the concentration gradient begins to relax, the coverage of oxygen remains close to zero, and the current drops. The overall charge associated with the process can be found by integrating the peak (current versus time). When the current is reversed, vacant sites begin to be oxidised back to oxygen covered sites as the potential approaches its open-circuit value and eventually the coverage of vacant sites approaches zero and a peaked response is seen.*

In the case of an irreversible charge-transfer process the rate of electron transfer is insufficient to maintain the charge-transfer process at equilibrium. The shape of the cyclic voltammogram is modified and peak positions shift as a function of scan rate (unlike the reversible case). A more detailed discussion can be found elsewhere.[93]

It must be remembered that in aqueous systems the redox process occurs over the entire electrode area, whereas in solid electrolyte systems the redox process occurs only in the three-phase or charge-transfer region. The technique has been used with solid electrolyte systems for sometime to study the oxidation and reduction of metals and metal oxides in inert atmospheres,[94,95] the behaviour of solid oxide fuel cell (SOFC) electrodes and has also been applied to the in-situ study of catalysts.[31,32,95]

Fabry and Kleitz[94] used cyclic voltammetry to study the behaviour of copper dissolved in zirconia in the temperature range of 1000 - 1200 K. More recently, van Manen and co-workers[95] have used cyclic voltammetry between 700 and 900°C to investigate the behaviour of a number of metal/metal oxide systems consisting of Fe/Fe_2O_3, Ni/NiO, Cu/CuO, Co/Co_2O_3. Ni/NiO was used as the reference electrode. All electrodes were prepared as tablets from physical mixtures of metal and metal oxide. All of the systems were found to be irreversible. Peaks in the cyclic voltammogrammes were ascribed to formation of oxide and metal layers which acted as diffusional barriers at the surface of the electrodes.

Vayenas and co-workers[31,32] used cyclic voltammetry to investigate oxygen adsorption and ethylene oxidation over a Pt catalyst-electrode. A cathodic

* It should be noted that the charge associated with the anodic and cathodic processes need not be equal as oxygen will have time to adsorb or desorb during the time scale of the cyclic voltammetry.

oxygen reduction peak was observed due to removal of oxygen from the three-phase (tp) region,

$$O(tp) + V_O^{\cdot\cdot} + 2e' + \Leftrightarrow O_O \quad (4.2)$$

Integrating the area of the peak gave the amount of oxygen adsorbed in the three-phase boundary region (this assumes no appreciable adsoprtion or surface diffusion during the process). This amount was a factor of 185 smaller than the total amount of oxygen adsorbed on the electrode at 399°C and an oxygen partial pressure of 21 kPa. While varying the oxygen partial pressure the amount of oxygen in the three-phase region was investigated. It was found that a Temkin isotherm could be used to describe the oxygen coverage in the three-phase region. However, the work appears to assume that the three-phase region is the same size in all cases, which must be an approximation as competing adsorption/desorption, surface diffusion and charge transfer would determine the size of the region. However, the concept is useful for providing an insight into the behaviour of the three-phase region.

When the area of the cathodic reduction peak was investigated as a function of temperature it was found to go through a maximum disappearing around 570°C. At low temperatures only a small amount of surface oxygen was available for reaction because of the low exchange current densities while, at higher temperatures, oxygen adsorption became limiting. At even higher temperatures thermodynamic equilibrium between oxygen adsorbed in the three-phase region and the gas phase was achieved. Eventually, the cathodic peak disappeared at a temperature close to the temperature of decomposition of PtO_2 as a result of the higher electrocatalytic activity of an oxide free surface.

At a temperature of 400°C the length of the three-phase line was estimated from the data on the coverage of oxygen in the three-phase region coupled with information on the rate of surface diffusion.[96] It was estimated that the thickness of the three-phase region was approximately 100 nm and that around 360 atoms of oxygen were reduced for every three-phase electrocatalytic site during the cathodic peak. The corresponding length of the three-phase line agreed well with SEM pictures of the electrode which appeared to be formed of around 1 μm crystallites.

The amount of oxygen adsorbed in the three-phase region was found to depend linearly with the exchange current density for different catalyst-electrodes under similar conditions. This indicates that the electrocatalytic reaction takes place at the three-phase boundary. It was pointed out that for less porous electrodes the charge-transfer reaction at the two-phase boundary might become important and that under some conditions oxygen on the electrolyte surface itself might play a role.

Cyclic voltammetry was also used under conditions of ethylene oxidation.[31] The rate of carbon dioxide production was seen to vary with the potential of the cell as would be expected from a system exhibiting NEMCA. Cyclic voltammetry was used to estimate the coverage of oxygen under working conditions by comparing the cathodic oxygen reduction peak with the peak obtained in the absence of reaction.

4.3 Other Techniques – Other electrochemical techniques that could be employed in sensor technology would include potential-step methods (or chronoamperometry, as current is recorded with time), current-step methods (or chronopotentiometry, as potential is recorded with time) and AC impedance. None of these techniques appear to have yet been applied to catalyst sensing in a systematic way.

Potential-step and current-step methods require existing electrochemical knowledge of any system if they are to be applied in sensing. Analysis of the results is complex as the cell response must be modelled based on pre-existing knowledge. Conversely, the technique of AC impedance is more straightforward. This involves applying a sinusoidal voltage to an electrochemical cell and measuring the response in terms of the variation in current. Generally, the current will have the same frequency as the applied voltage, but will differ in amplitude and phase. Typically, the frequency is varied between 10^{-2} and 10^5 Hz. For analysis purposes it is necessary to operate in a region where there is a linear dependence of rates upon electrode potential. Therefore, the peak-to-peak amplitude of the voltage should not exceed 10 mV so to remain in the linear region of the Butler-Volmer equation (this is beneficial for sensing in that the surface processes are unlikely to be disturbed). An AC analyser is used to measure the amplitude of the alternating current and its phase angle relative to the applied voltage. The real and imaginary parts of the impedance are then displayed as a function of frequency in the complex plane. Essentially, the technique separates processes with different time constants; electrolyte grain boundary, electrolyte bulk and interface effects all yield separate semicircles in the complex plane (additionally, in the presence of a diffusional limitation a straight line of unit slope is obtained at low frequencies). Ideally, AC impedance would yield the resistance and capacitance associated with the individual steps of the overall electrode process. However, the resistance associated with charge transfer or mass transfer is more readily obtained in the form of the exchange current density or limiting current. The double-layer capacitance is also determined, although it is difficult to know how this is relevant to the catalytic processes occurring on the gas-exposed surface of the catalyst.

5 Summary and Future Opportunities

The technique of solid electrolyte potentiometry has been widely applied to a number of oxidation reactions over metal catalysts for measuring the thermodynamic activity of oxygen on a metal surface under reaction conditions. However, care must be taken when interpreting results using the Nernst equation as mixed potentials may occur in the presence of strongly reducing reactants such as carbon monoxide. In particular, the technique has been successful for studying the phenomenon of periodic reaction rate oscillations and has been used to demonstrate that such oscillations are often related to the formation and destruction of an oxide over-layer. More recently, the technique has been applied to the study of oxide catalysts. Although the interpretation of the e.m.f. can be more complex for oxide catalysts, it is envisaged that the use of SEP will continue to expand in this area because of the importance of oxide catalysts for oxidation processes in general and partial oxidation processes in particular.

Amperometric techniques can also be useful in gaining information on the state of a catalyst. These techniques tend to be more complex than potentiometric methods and are as a result less developed. In particular, early work with the technique of cyclic voltammetry has produced some promising results.

Up until now the use of solid electrolyte cells has been confined to laboratory studies. However, there is no reason why cells could not be adapted to sense the state of a catalyst in a working heterogeneous catalytic reactor.[97] The catalyst-electrode would be of the same or similar material to the catalyst in the rest of the reactor for the most relevant information. There is a good deal of scope for developing such technology and many crucial points would need to be addressed. For instance, how similar does electrode composition need to be to the catalyst composition? What is the influence of electrode morphology? How does sensor response change during catalyst deactivation? The benefits from appropriate sensors could be particularly significant for reactions such as partial oxidation processes. It is well known that the state of an oxide catalyst can influence activity and selectivity for a partial oxidation process. If the catalyst is too oxidised selectivity is poor; if the catalyst is too reduced activity is poor. Consequently, there may be an optimum catalyst state for partial oxidation product yield. Any sensor that could be employed to give information regarding catalyst state would therefore be of great potential benefit.

Sensors could easily be miniaturised for such applications. During laboratory experiments it is most common to use a reference electrode exposed to air. This would be inconvenient if the sensor were to be used in a reactor. Instead the sensor could be miniaturised by using a mixed metal/metal oxide reference electrode. The reference oxide would be depleted under reducing conditions because of electronic leakage across the electrolyte but should be easily regenerated with the application of an appropriate potential.

In conclusion, a good deal of fundamental work has been performed with solid electrolyte electrochemical cells on sensing the state of working catalysts. Studies on oxide catalysts have only recently commenced and there remain many other systems yet to be investigated. However, this technology has not yet been seriously applied to challenges such as improving reactor control. There remains much scope for progress in such sensor research and development.

Nomenclature

E	electromotive force or e.m.f. (V)
F	faraday constant (9.65×10^4 C mol^{-1})
I_i	current associated with ith species (A)
i	current density (A m^{-2})
i_0	exchange current density (A m^{-2})
i_l	limiting current density (A m^{-2})

IR	resistive loss in a cell (V)
k_A	rate constant for the anodic branch of a reaction (mol s^{-1} m^{-2})
k_C	rate constant for the cathodic branch of a reaction (mol s^{-1} m^{-2})
P_{O2}	partial pressure of oxygen (bar)
R	gas constant (8.314 J mol^{-1} K^{-1})
R_A	rate of anodic reaction (mol s^{-1} m^{-2})
R_C	rate of cathodic reaction (mol s^{-1} m^{-2})
R_{ct}	resistance due to charge-transfer (Ω m^2)
R_{mt}	resistance due to mass transfer (Ω m^2)
t_{ion}	ionic transfer number (dimensionless)
T	temperature (K)
V	voltage (V)
x	distance co-ordinate (m)
z_i	charge on *i*th species
α_O	thermodynamic activity of adsorbed oxygen ($bar^{0.5}$)
β	symmetry factor (dimensionless)
$\Delta\phi$	potential difference (V)
ϕ	electrostatic potential (V)
γ	constant dependent upon conditions at counter electrode (dimensionless)
η	electrode overpotential (V)
μ_i	chemical potential of *i*th species (J mol^{-1})
$\bar{\mu}_i$	electrochemical potential of *i*th species (J mol^{-1})
θ_i	coverage of *i*th species (dimensionless)
σ_i	conductivity of *i*th species (A m V^{-1})

Subscripts

A	denotes the anode
C	denotes the cathode or counter electrode
ct	denotes charge transfer
i	denotes *i*th species
mt	denotes mass transfer
R	denotes the reference electrode
W	denotes the working electrode

Superscripts

0	denotes equilibrium

References

1. Geller, S. (ed.), 'Solid Electrolytes', Springer-Verlag, Berlin-Heidelberg, 1977.
2. Subbarao, E.C., 'Solid Electrolytes and their Applications', Plenum Press, New York, 1990.
3. Kröger, F.A.; Vink, H.J., in 'Solid State Physics', Vol. 3, p307; Seitz, F.; Turnbull, D. (eds), Plenum Press, New York, 1956.
4. Maskell, W.C.; Steele, B.C.H., *J. Appl. Electrochem.* **16** (1986) 475.
5. Vayenas, C.G.; Bebelis, S.; Yentekakis, I.V.; Lintz, H.-G., *Catalysis Today* **11** (1992) 303-442.

6. Vayenas, C.G., *Solid State Ionics* **28-30** (1988) 1521.
7. Lintz, H.-G.; Vayenas, C.G., *Angew. Chem. Int. Ed. Engl.* **28** (1989) 708-15.
8. Gellings, P.J.; Koopmans, H.J.A.; Burggraaf, A.J., *Appl. Catal.* **39** (1988) 1.
9. Stoukides, M., *Ind. Eng. Chem. Res.* **27** (1988) 1745.
10. Metcalfe, I.S., *Catal. Today* **20** (1994) 283-294.
11. Vayenas, C.G.; Bebelis, S.; Yentekakis, I.V.; Tsiakaris, P.; Karasali, H., *Platinum Metals Rev.* **34** (1990) 122-130.
12. Vayenas, C.G.; Bebelis, S.; Yentekakis, I.V.; Neophytides, S., *Solid State Ionics* **53-56** (1992) 97-110.
13. Wagner, C., *Z. Phys. Chem.* **B21** (1933) 25.
14. Weissbart, J.; Ruka, R., *Rev. Sci. Instr.* 32 (1961) 593.
15. Metcalfe, I.S.; Sundaresan, S., *AIChE. J.* **34** (1988) 195-208.
16. Gessner, M.A.; Nagy, S.G.; Michaels, J.M., *J. Electrochem. Soc.* **135** (1988) 1294-1301.
17. Mahmood, M.N.; Bonanos, N., *Solid State Ionics* **53-56** (1992) 142-148.
18. Agrawal, Y.K.; Short, D.W.; Gruenke, R.; Rapp, R.A., *J. Electrochem. Soc.* **121** (1974) 354.
19. Alcock, C.B.; Zador, S., *J. Appl. Electrochem.* 2 (1972) 289.
20. Yuan, D.; Kröger, F.A., *J. Electrochem. Soc.* 116 (1969) 594.
21. Bard, A.J.; Faulkner, L.R., 'Electrochemical Methods: Fundamentals and Applications', John Wiley, New York, 1980.
22. Göpel, W., Proc. of Intl. Symp. on SOFC's, Nagoya, Japan, November, 1989.
23. Schindler, K.; Schmeisser, D.; Vohrer, U.; Wiemhofer, H.-D.; Göpel, W., *Sensors and Actuators* **17** (1989) 377.
24. Wang, D.Y.; Nowick, A.S., *J. Electrochem. Soc.* **126** (1979) 1155.
25. Wang, D.Y.; Nowick, A.S., *J. Electrochem. Soc.* **126** (1979) 1166.
26. Wang, D.Y.; Nowick, A.S., *J. Electrochem. Soc.* **127** (1980) 113.
27. Wang, D.Y.; Nowick, A.S., *J. Electrochem. Soc.* **128** (1981) 55.
28. Robertson, N.L.; Michaels, J.N., *J. Electrochem. Soc.* **137** (1990) 129-135.
29. Mizusaki, J.; Amano, K.; Yamauchi, S.; Fueki, K., *Solid State Ionics* **22** (1987) 313.
30. Nguyen, B.C.; Rincon-Rubio, L.M.; Mason, D.M., *J. Electrochem. Soc.* **133** (1986) 1807-1860.
31. Vayenas, C.G.; Ioannides, A.; Bebelis, S., *J. Catal.* **129** (1991) 67-87.
32. Yi, J.; Kaloyannis, A.; Vayenas, C.G., *Electrochimica Acta* **38** (1993) 2533-2539.
33. Mari, C.M., *Solid State Ionics* **12** (1984) 419-430.
34. Burggraaf, A.J.; van Dijk, M.P.; de Vries, K.J., *Solid State Ionics* **18/19** (1986) 807-812.
35. Petrolekas, P.; Metcalfe, I.S., *J. Catal.* **152** (1995) 147-163.
36. Mizusaki, J.; Tagawa, H., Proc. Intl. Symp. on SOFC's, Nagoya, Japan, November 1989.
37. Mizusaki, J.; Tagawa, H.; Tsuneyoshi, K.; Sawata, A., Proc. SOFC Meeting, pp 254-265, Hollywood, Florida, 1989.
38. Wagner, C., *Adv. Catal.* **21** (1970) 323-381.
39. Vayenas, C.G.; Saltsburg, H., *J. Catal.* **57** (1979) 296-309.
40. Saranteas, C.; Stoukides, M., *J. Catal.* **93** (1985) 417-429.
41. Eng,D.; Stoukides, M.; McNally, T., *J. Catal.* **106** (1987) 342-353.
42. Arif, H.; Stoukides, M., *Chem. Eng. Sci.* **41** (1986) 945-952.
43. Hetrick, R.E.; Logothetis, E.M., *Appl. Phys. Lett.* **34** (1979) 117.
44. Haaland, D.M., *J. Electrochem. Soc.* **127** (1980) 796.
45. Okamoto, H.; Obayashi, H.; Kudo, T., *Solid State Ionics* **1** (1980) 319.
46. Okamoto, H.; Kawamura, G.; Kudo, T., *Solid State Ionics* **3/4** (1981) 453-456.
47. Okamoto, H.; Obayashi, H.; Kudo, T., *Solid State Ionics* **3** (1981) 453.
48. Okamoto, H.; Kawamura, G.; Kudo, T., *J. Catal.* **82** (1983) 332.
49. Okamoto, H.; Kawamura, G.; Kudo, T., *J. Catal.* **82** (1983) 322.
50. Okamoto, H.; Kawamura, G.; Kudo, T., *J. Catal.* **90** (1984) 374.
51. Okamoto, H.; Kawamura, G.; Kudo, T., *J. Catal.* **87** (1984) 1-9.

52. Okamoto, H.; Kawamura, G.; Kudo, T., *J. Catal.* **86** (1984) 437.
53. Engel, T.; Ertl, G., *Adv. Catal.* **28** (1979) 2.
54. Vayenas, C.G., *J. Catal.* **90** (1984) 371.
55. Mizusaki, J.; Tagawa, H.; Miyaki, Y.; Yamauchi, S.; Fueki, K.; Koshiro, I.; Hirano, K., *Solid State Ionics* **53-56** (1992) 126-134.
56. Ehrhardt, J.J.; Häfele, E.; Lintz, H.-G.; Martins, A.F., *Ber. Bunsenges. Phys.-Chem.* **89** (1985) 894-6.
57. Häfele, E.; Lintz, H.-G., *Solid State Ionics* **23** (1987) 235-239.
58. Yentekakis, I.V.; Neophytides, S.; Vayenas, C.G., *J. Catal.* **111** (1988) 152.
59. Turner, J.E.; Sales, B.C.; Maple, M.B., *Surf. Sci.* **103** (1981) 54.
60. Sales, B.C.; Turner, J.E.; Maple, M.B., *Surf. Sci.* **114** (1982) 381.
61. Sandler, Y.L., *J. Electrochem. Soc.* **118** (1971) 1378.
62. Seimanides, S.; Stoukides, M., *J. Catal.* **88** (1984) 490.
63. Seimanides, S.; Stoukides, M., *J. Catal.* **98** (1986) 540.
64. Mahmood, M.N.; Bonanos, N., *Solid State Ionics* **53-56** (1992) 142-248.
65. Vayenas, C.G.; Lee, B.; Michaels, J.N., *J. Catal.* **66** (1980) 36.
66. Vayenas, C.G.; Georgakis, C.; Michaels, J.N.; Tormo, J., *J. Catal.* **67** (1981) 348-360.
67. Stoukides, M.; Vayenas, C.G., *J. Catal.* **64** (1980) 18.
68. Eng, D.; Stoukides, M., *J. Catal.* **130** (1991) 306-309.
69. Hillary, A.; Stoukides, M., *J. Catal.* **113** (19880 295.
70. Stoukides, M.; Vayenas, C.G., *J. Catal.* **69** (1981) 18.
71. Stoukides, M.; Vayenas, C.G., *J. Catal.* **82** (1983) 45-55.
72. Stoukides, M.; Vayenas, C.G., *J.Electrochem. Soc.* **131** (1984) 839-45.
73. Stoukides, M.; Vayenas, C.G., 8th ICC DECHEMA **4** (1984) 827-833.
74. Stoukides, M.; Vayenas, C.G., *J. Catal.* **74** (1982) 266-275.
75. Freriks, I.C.; Bouwman, R.; Geenen, P.V., *J. Catal.* **65** (1980) 311.
76. Häfele, E., Dissertation, Universität Karlsruhe, 1987.
77. Häfele, E.; Lintz, H.-G., *Ber. Bunsenges. Phys.-Chem.* **92** (1988) 188.
78. Lintz, H.-G.; Oerter, M., *Platinum Metals Rev.* **37** (1993) 8-13.
79. Bruck, J.; Lintz, H.G.; Oerter, M., *Catal. Today* **17** (1993) 95-102.
80. Vayenas, C.G.; Michaels, J.N., *Surf. Sci.* **120** (1982) L405.
81. Berry, R.J., *Surf. Sci.* **76** (1978) 415.
82. Okamoto, H.; Kawamura, G.; Kudo, T., *Electrochim Acta* **28** (1983) 379.
83. Balian, A,; Hatzigiannis, G.; Eng, D.; Stoukides, M., *J. Catal.* **145** (1994) 526-536.
84. Greger, M.; Ihme, B.; Kotter, M.; Riekert, L., *Ber. Bunsenges. Phys. Chem.* **88** (1984) 427.
85. Greger, M.; Riekert, L., *Ber. Bunsenges. Phys. Chem.* **91** (1987) 1007.
86. Breckner, E.M.; Sundaresan, S.; Benziger, J.B., *Appl. Catal.* **30** (1987) 277.
87. Hildenbrand, H.-H.; Lintz, H.-G., *Appl. Catal.* **65** (1990) 241-252.
88. Hildenbrand, H.-H.; Lintz, H.-G., *Catalysis Today* **9** (1991) 153-160.
89. Greger, M., Thesis, University of Karlsruhe, 1987.
90. Pizzini, S.; Mari, C.M.; Hadjicostantis, D.C., in 'Reactivity of Solids', Wood, J.; Lindquist, O.; Helgesson, C.; Vannerberg, N.-G. (eds), Proc. 8th Intl. Symp. on Reactivity of Solids, Plenum Press, New York, 1977, p161.
91. Pizzini, S.; Mari, C.M.; Zanderighi, L., *Gazz. Chim. Ital.* **110** (1980) 389.
92. Alqahtany, H.; Eng, D.; Stoukides, M., *J. Electrochem. Soc.* **140** (1993) 1677-1681.
93. Greef, R.; Peat, R.; Peter, L.M.; Pletcher, D.; Robinson, J., "Instrumental Methods in Electrochemistry", Southampton Electrochemistry Group, 1985.
94. Fabry, P.; Kleitz, M., *J. Electrochem. Soc.* **126** (1979) 2183-7.
95. van Manen, P.A.; Weewer, R.; de Wit, J.H.W., *J. Electrochem. Soc.* **139** (1992) 1130.
96. Lewis, R.; Gomer, R., *Surf. Sci.* **12** (1968) 157.
97. Petrolekas, P.; Metcalfe, I.S., *J. Electrochem. Soc.* **142** (1995) 952-957.

2

Molecular Engineering of Supported Metal Oxide Catalysts: Oxidation Reactions over Supported Vanadia Catalysts*

BY ISRAEL E. WACHS

1 Introduction

Supported metal oxide catalysts consist of an active metal oxide component (e.g., oxides of V, Mo, Cr, W, Re, etc.) deposited on the surface of an oxide support (e.g., Al_2O_3, SiO_2, TiO_2, etc.). In certain instances, the supported metal oxide catalysts are also combined with noble metals in order to enhance the catalytic properties of the noble metal component (e.g., $Pt/CeO_2/Al_2O_3$ and $Pt/SnO_2/Al_2O_3$). Such catalysts are extensively employed in the chemical, petroleum and pollution control industries.[1-4] The industrial development of supported metal oxide catalysts over the past five decades is listed chronologically in Table 1. The V_2O_5-$K_2S_2O_7/SiO_2$ catalyst employed for sulfuric acid manufacture since ~1918, via the oxidation of SO_2 to SO_3, is intentionally not listed in Table 1 because, unlike the other supported metal oxide catalysts, the active vanadia component is present as a molten salt under reaction conditions (supported liquid phase catalyst). The initial industrial applications of supported metal oxide catalysts were limited to hydrocarbon dehydrogenation/hydrogenation and olefin polymerization/metathesis reactions. In more recent years, the number of applications of supported metal oxide catalysts for oxidation reactions has grown significantly due to their excellent oxidation characteristics in the manufacture of certain chemical intermediates and pollution control strategies. Among the supported metal oxide catalysts employed for oxidation reactions, supported vanadia catalysts have emerged as the most versatile oxidation catalysts.[5] For example, supported vanadia catalysts are industrially used to oxidize o-xylene to phthalic anhydride (V_2O_5/TiO_2),[6,7] ammoxidation of alkyl aromatics to aromatic nitriles (V_2O_5/Al_2O_3 and V_2O_5/TiO_2)[7] and the selective catalytic reduction (SCR) of NO_x emissions with NH_3 to N_2 ($V_2O_5/WO_3/TiO_2$ and $V_2O_5/MoO_3/TiO_2$)[8]. In conventional SCR applications, the oxidation of the SO_2 in the flue gas to SO_3 is minimized because of the potential formation of deleterious ammonium sulfates, but in the DESONOX process the oxidation of SO_2 to SO_3 is desired for the simultaneous manufacture of sulfuric acid ($V_2O_5/WO_3/TiO_2$)[9]. In addition to the current commercial applications,

* Based on Seminar Presented at the 13th Award for Excellence in Catalysis Lecture, sponsored by Exxon Research and Engineering Company, The Catalysis Society of Metropolitan New York Meeting, Somerset, NJ (May 15, 1996).

supported vanadia catalysts are also capable of performing many other oxidation reactions: oxidation of methane to formaldehyde (V_2O_5/SiO_2),[10] oxidation of ethane to ethylene (V_2O_5/SiO_2),[11] oxidation of propane to propylene (V_2O_5/Nb_2O_5),[12-14] oxidation of butane to butenes,[15,16] oxidation of butane to maleic anhydride ($V_2O_5/P_2O_5/TiO_2$, $V_2O_5/P_2O_5/Al_2O_3$, and $V_2O_5/P_2O_5/SiO_2$)[17-20] and pentane oxidation to maleic anhydride and phthalic anhydride.[21]

The present chapter will primarily focus on oxidation reactions over supported vanadia catalysts because of the widespread applications of these interesting catalytic materials.[5,6,22-24] Although this article is limited to well-defined supported vanadia catalysts, the supported vanadia catalysts are model catalyst systems that are also representative of other supported metal oxide catalysts employed in oxidation reactions (e.g., Mo, Cr, Re, etc.).[25,26] The key chemical probe reaction to be employed in this chapter will be methanol oxidation to formaldehyde, but other oxidation reactions will also be discussed (methane oxidation to formaldehyde, propane oxidation to propylene, butane oxidation to maleic anhydride, CO oxidation to CO_2, SO_2 oxidation to SO_3 and the selective catalytic reduction of NO_x with NH_3 to N_2 and H_2O). This chapter will combine the molecular structural and reactivity information of well-defined supported vanadia catalysts in order to develop the molecular structure-reactivity relationships for these oxidation catalysts. The molecular structure-reactivity relationships represent the molecular ingredients required for the molecular engineering of supported metal oxide catalysts.

Table 1 *Industrial Development of Supported Metal Oxide Catalysis*

1940's –	CrO_3/Al_2O_3	n-butane dehydrogenation to butene
1950's –	CrO_3/SiO_2	ethylene polymerization
1960's –	MO_3/Al_2O_3	hydrodesulfurization (HDS) of Crude Oil
–	WO_3/Al_2O_3	hydrodesulfurization (HDS) of Crude Oil
–	MoO_3/Al_2O_3	higher n-alkane dehydrogenation
	V_2O_5/TiO_2	oxidation of o-xylene to phthalic anhydride
1970's –	TiO_2/SiO_2	epoxidation of propylene by hydrogen peroxide
–	Re_2O_7/Al_2O_3	olefin metathesis
–	$V_2O_5/WO_3/TiO_2$	selective catalytic reduction (SCR) of NO_x emissions
–	$Pt/Pd/Rh/CeO_2/Al_2O_3$	automotive pollution control
1980's –	$Pt/SnO_2/Al_2O_3$	gasoline reforming
–	V_2O_5/TiO_2	ammoxidation of alkyl aromatics
–	$Fe_2O_3/CrO_3/Al_2O_3$	oxidation of H_2S to elemental sulfur (Super Claus)
1990's –	$CuO/CrO_3/Al_2O_3$	oxidation of VOCs to Cl-VOCs
–	V_2O_5/TiO_2	oxidation of Dioxin and PCB emissions
–	$V_2O_5/MoO_3/TiO_2$	selective catalytic reduction (SCR) of NO_x emissions
–	Fe_2O_3/SiO_2	oxidation of H_2S to elemental sulfur (Super Claus)
–	NiO/ZrO_2	N_2O decomposition to N_2 and O_2
–	SO_4/ZrO_2	solic acid catalyst
–	WO_3/ZrO_2	solic acid catalyst

2 Molecular Characterization

The catalysis science of supported metal oxide catalysts, especially supported vanadia catalysts, has lagged behind their industrial development.[25,26] In the 1970s, two models were proposed for the active metal oxide component: a three-dimensional microcrystalline phase (e.g., small metal oxide crystallites) or a two-dimensional surface metal oxide overlayer (e.g., surface metal oxide monolayer). In the 1980s, many studies demonstrated that the active metal oxide components were primarily present as two-dimensional surface metal oxide overlayers, below monolayer coverage, and that the surface metal oxide overlayers control the catalytic properties of supported metal oxide catalysts. The synergistic interaction between the surface vanadia overlayer and the underlying oxide support prompted Gellings to state ". . . that neither the problem of the structure of suppored vanadium oxide nor that of the special role of TiO_2 as a support have definitely been solved. Further work on these and related topics is certainly necessary."[22] In more recent years, many fundamental studies have focused on the molecular structural determination of the surface vanadia phase and to a lesser extent the molecular structure-reactivity relationships of supported vanadia catalysts.[27-30]

The molecular structures of the active surface vanadia species have been elucidated with the use of multiple molecular spectroscopies: X-ray absorption spectroscopy (XANES/EXAFS),[31] solid state nuclear magnetic resonance (NMR),[32,33] Raman,[26,33-37] IR[26,37] and oxygen-18 exchange experiments.[30,37] These combined characterization studies revealed that, at elevated temperatures and under oxidizing conditions or dehydrated conditions at room temperature, the surface vanadia species was present as both isolated and polymerized VO_4 units containing one terminal V=O bond and three bridging V-O-M bonds (where M=V or the support cations). For silica-supported vanadia catalysts, however, only isolated surface VO_4 units were present because of the inability to achieve high surface vanadia coverages on this somewhat unreactive surface.[33] Thus, the active surface vanadia species in supported metal oxide catalysts possess terminal V=O, bridging V-O-V and bridging V-O-Support bonds.

3 Which Metal Oxide Bonds Are Critical for Oxidation Reactions: V=O, V-O-V or V-O-Support?

3.1 Bridging V-O-V Bonds. – Several recent *in situ* Raman studies have demonstrated that the ratio of bridging V-O-V bonds to terminal V=O bonds increases with surface vanadia coverage on oxide supports, with the exception of V_2O_5/SiO_2 which only possesses isolated surface vanadia species and no bridging V-O-V bonds.[33-36] Consequently, the role of bridging V-O-V bonds in oxidation reactions can be chemically probed by examining the reaction turnover frequency, TOF - number of molecules reacted per surface vanadia site per second, as a function of surface vanadia coverage. The numbers of surface vanadia sites were determined from the amount of vanadia present in the catalysts since

Raman spectroscopy demonstrated 100% dispersion of the supported vanadia phase. The number of molecules reacted was determined by the conversion of the reactants, under differential reaction conditions and the absence of heat and mass transfer limitations, in fixed-bed reactor studies. The TOFs for several oxidation reactions over V_2O_5/TiO_2 catalysts, monolayer coverage for this TiO_2 support (Degussa P-25, ~55 m^2/g) corresponds to ~6% V_2O_5/TiO_2, are presented in Figure 1.

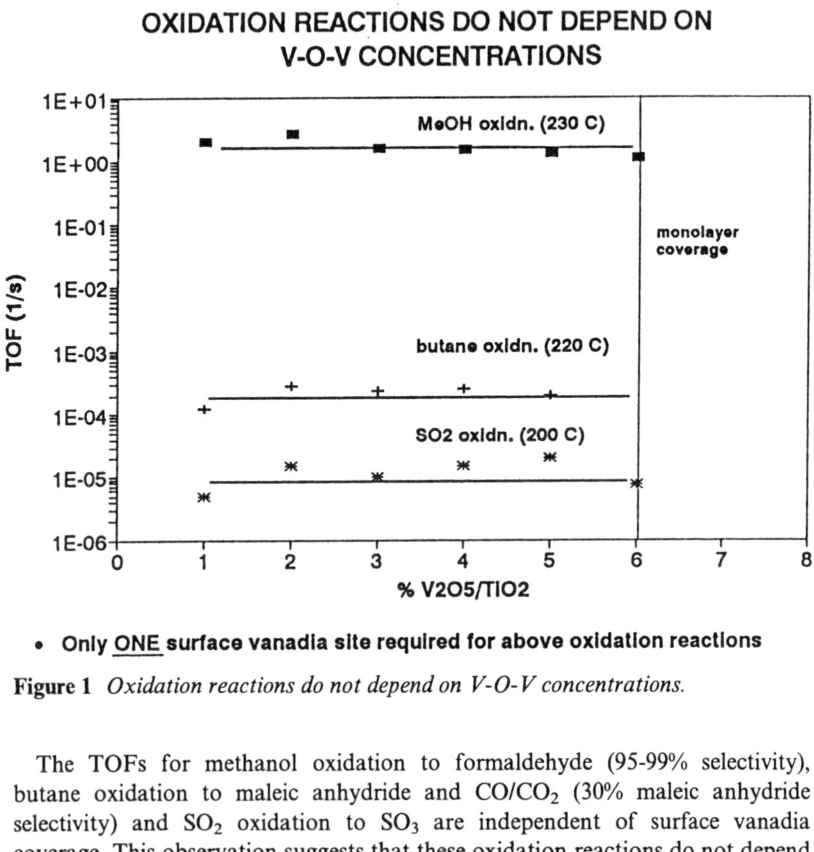

Figure 1 *Oxidation reactions do not depend on V-O-V concentrations.*

The TOFs for methanol oxidation to formaldehyde (95-99% selectivity), butane oxidation to maleic anhydride and CO/CO_2 (30% maleic anhydride selectivity) and SO_2 oxidation to SO_3 are independent of surface vanadia coverage. This observation suggests that these oxidation reactions do not depend on the surface concentration of bridging V-O-V bonds since the reaction TOFs do not correlate with the surface density of bridging V-O-V bonds. Furthermore, the constant TOFs with surface vanadia coverage suggest that only one surface vanadia site is required for the activation of these molecules during the oxidation reactions.

3.2 Terminal V=O Bonds. – The terminal V=O bonds can also be directly monitored with *in situ* Raman spectroscopy during the oxidation reactions, and the vibrational frequency of the terminal V=O bond is directly related to its bond strength (stronger or shorter bonds vibrate at high cm^{-1} and weaker or longer bonds vibrate at lower cm^{-1}). *In situ* Raman measurements during butane oxidation over several vanadia supported catalysts (V_2O_5/TiO_2, V_2O_5/ZrO_2 and V_2O_5/Al_2O_3) revealed essentially the same terminal V=O bond strength for all the catalysts, 1025-1028 cm^{-1}, but the butane oxidation TOFs varied by more than an order of magnitude.[20] Consequently, the butane oxidation TOFs do not correlate with the characteristics of the terminal V=O bond, which suggests that this bond is not critical to this oxidation reaction. A similar lack of correlation between the terminal V=O bond characteristics and the methanol oxidation TOFs was also previously found for V_2O_5/TiO_2, V_2O_5/ZrO_2, V_2O_5/Nb_2O_5, V_2O_5/Al_2O_3 and V_2O_5/SiO_2 catalysts.[29] Additional support for this conclusion comes from oxygen-18 labeled experiments of the terminal V=O bond.[38] In these experiments, the oxygen-18 label was introduced into the surface vanadia species prior to initiation of the butane oxidation reaction with oxygen-16. The exchange rate of the oxygen-18 labeled terminal V=O bond was monitored with *in situ* Raman spectroscopy during butane oxidation. The experiment demonstrated that the terminal $V={}^{18}O$ bond was very stable and that ~30 minutes was required to completely exchange the oxygen-18 with oxygen-16 in the terminal V=O bond. This corresponds to an oxygen exchange rate that is approximately 20 times longer than the characteristic reaction time. Thus, the *in situ* Raman spectroscopy experiments during oxidation reactions over supported vanadia catalysts suggest that the terminal V=O bond is not critical to oxidation reactions.

3.3 Bridging V-O-Support Bonds. – Unlike the terminal V=O bonds and the bridging V-O-V bonds, the bridging V-O-Support bond can not be directly monitored with Raman spectroscopy because it is not Raman active due to its slightly ionic character.[26] IR spectroscopy, unfortunately, also cannot detect this band because of strong IR absorption by the oxide supports in the region of interest (~600-700 cm^{-1})[26]. The characteristics of the bridging V-O-Support bond, however, can be altered by varying the specific oxide support or the oxide support ligand (e.g., oxides of Ce, Zr, Ti, Nb, Al and Si). The methanol oxidation TOFs over a series of supported vanadia catalysts is shown in Figure 2.

The methanol oxidation TOFs vary by approximately four orders of magnitude with the specific oxide support ($CeO_2 > ZrO_2 \sim TiO_2 > Nb_2O_5 > Al_2O_3 > SiO_2$)[29]. This strong influence of the oxide support suggests that the bridging V-O-Support bond controls the specific catalytic activity or TOF. In Figure 2, the methanol oxidation TOFs are correlated with the Sanderson electronegativities of the oxide support cations. This correlation suggests that a higher electron density of the bridging oxygen in the V-O-Support bond, corresponding to lower electronegativity of the oxide support cation, enhances the methanol oxidation TOF. It is necessary to examine the mechanism and kinetics of methanol oxidation in order to fully understand the influence of the oxide support ligands upon the reaction TOFs.

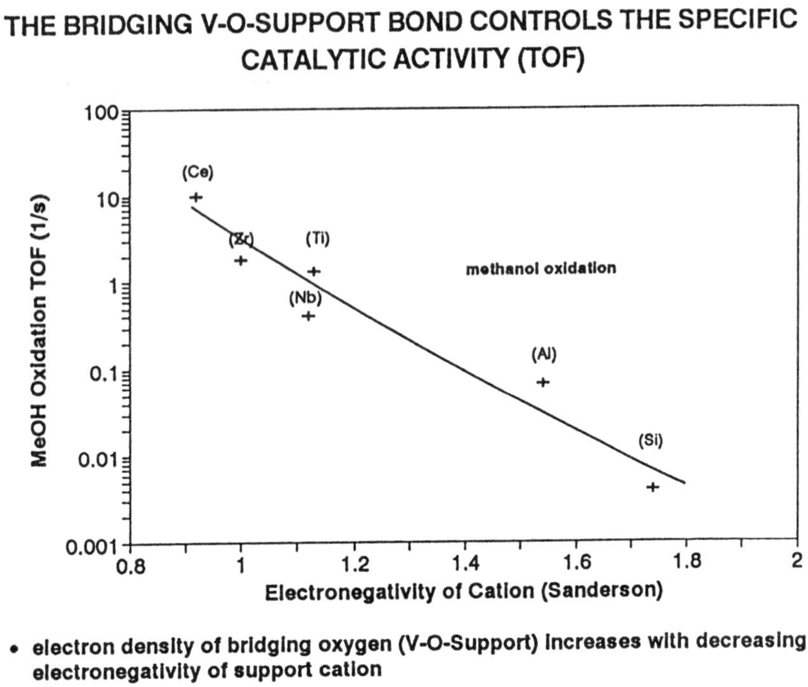

Figure 2 *The bridging V-O-Support bond controls the specific catalytic activity.*

4 Methanol Oxidation Reaction Kinetics and Influence of the Oxide Support

The mechanism and kinetics of methanol oxidation over oxide supports have been extensively examined in recent years.[39,40] It has been shown that methanol reversibly interacts with oxide supports by forming a surface methoxy species, CH_3O, and a surface hydroxyl group, OH, and that the rate determining step is breaking of a methyl C-H bond to form formaldehyde. Thus, the kinetics of methanol oxidation can be represented by an equilibrated adsorption step, where K is the equilibrium adsorption constant, and a rate determining reaction step, where k is the Arrhenius rate constant. This results in the following kinetic expression in the presence of excess oxygen and low concentrations of methanol, where $[CH_3OH]$ represents the partial pressure of methanol:

$$r = K\,k\,[CH_3OH] \quad (1)$$

$$K = A_1 \exp(-H/RT) \quad (2)$$

$$k = A_2 \exp(-E/RT) \quad (3)$$

where A_1 and A_2 are pre-exponential factors, and E and H are the reaction activation energy and the methanol heat of adsorption (an exothermic value), respectively. When the inhibition of water is also considered in the kinetic expression for an oxidized surface,[39] the equilibrium adsorption constant becomes a function of both the adsorption of methanol and the adsorption of water:

$$K = K_{methanol} / (K_{water})^{1/2} \tag{4}$$

and the reaction rate is modified by $1/[H_2O]^{1/2}$. The above kinetic rate expression, equation (1), suggests that the variations in the TOF of the different supported vanadia catalysts may be due to either changes in (1) the surface reaction rate constant, k, (2) the equilibrium adsorption constant, K, or (3) simultaneous changes in both the values of K and k.

In order to discriminate between these different possibilities, methanol oxidation temperature programmed reaction spectroscopy (TPRS) studies were undertaken over the series of supported vanadia catalysts since such experiments directly measure the fundamental surface reaction rates.[41] The TPRS experiments were conducted by adsorbing methanol on the supported vanadia catalysts at 100 °C, corresponding to monolayer coverage, and then increasing the sample temperature to monitor the decomposition of the surface methoxy species to formaldehyde via a mass spectrometer. The methanol oxidation TPRS experiments revealed that the rate constants, k, for the decomposition of the surface methoxy intermediate to formaldehyde were identical, decomposition temperature of ~210 °C, over the series of supported vanadia catalysts (with the exception of the V_2O_5/SiO_2 system to be discussed below). This suggests that the difference in the methanol oxidation TOFs among the supported vanadia catalysts is primarily related to the difference in the equilibrium adsorption constant, K.

Combination of this new insight about the role of the equilibrium adsorption constant with the electronegativity trend of the specific oxide support ligand, see Figure 2, suggests the methanol oxidation catalytic cycle shown in Figure 3: The adsorption of methanol occurs at the bridging V-O-S bond by protonation of the bridging oxygen and formation of a V-OCH_3 intermediate. The efficiency of this dissociative adsorption step, reflected in the magnitude of the equilibrium adsorption constant, is dependent on the stability of the bridging V-O-S bond. A higher electron density on the bridging oxygen, corresponding to a lower electronegativity of the oxide support ligand, results in a more basic bridging oxygen that enhances the adsorption of the mildly acidic methanol molecule. Conversely, a lower electron density on the bridging oxygen, corresponding to a higher electronegativity of the oxide support ligand, results in a less basic bridging oxygen that depresses the adsorption of the mildly acidic methanol molecule. Thus, the different TOFs for methanol oxidation over the series of supported vanadia catalysts should be related to the concentration of surface methoxy species during methanol oxidation. Preliminary *in situ* IR experiments were undertaken during methanol oxidation to directly measure the concentration of surface methoxy species on the supported vanadia catalysts, and the initial results appear to be consistent with this hypothesis.[42] The decomposition of the surface methoxy intermediate to formaldehyde probably only occurs on the surface vanadia site since the same surface reaction rate constant, k, is reflected in the

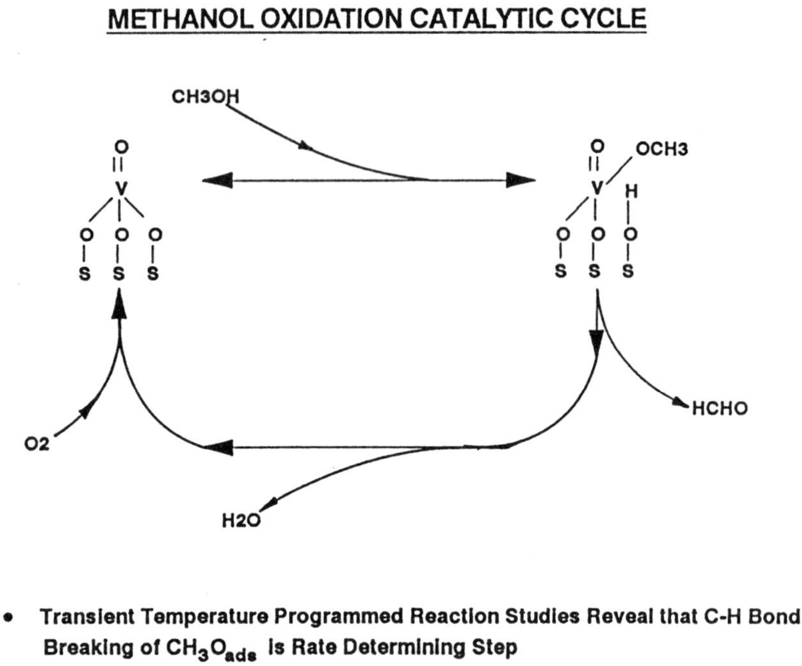

Figure 3 *Methanol oxidation catalytic cycle.*

TPRS experiments for this reaction step. The two hydrogens released in the methanol adsorption and the surface methoxy decomposition steps are eventually converted to water. Spectroscopic details about the formation of water are presently not available, but the formation of water most likely proceeds via the condensation of two surface hydroxyl groups. The reduced surface vanadia site is readily reoxidized back to vanadium (+5) by gas phase oxygen as shown by *in situ* Raman measurements.[42]

The V_2O_5/SiO_2 system, however, did not follow the above trend. The surface methoxy intermediate was significantly more stable on this catalyst, decomposition temperature of ~300 °C compared to ~210 °C for the other supported vanadia catalysts. This suggests that the surface reaction rate constant, k, for surface methoxy decomposition to formaldehyde is significantly lower for the silica supported vanadia catalyst compared to the other supported vanadia catalysts. Consequently, the low TOF for methanol oxidation over the silica supported vanadia catalyst is due to both a lower equilibrium adsorption constant, due to the high electronegativity of the silica ligand, and a lower surface reaction constant, k.

The apparent activation energies for methanol oxidation over the series of support vanadia catalysts is a composite of the activation energy and the heats of adsorption:

$$E_{app} = E + H_{methanol} - 1/2\, H_{water} \quad (5)$$

Fundamental TPRS studies have shown that the activation energy for breaking the C-H bond of methoxy is approximately ~23 kcal/mol.[39] The similar apparent activation energies for methanol oxidation over all the supported vanadia catalysts,[29] 19-23 kcal/mol, suggests that the apparent heats of adsorption,

$$H_{app} = H_{methanol} - 1/2\, H_{water} \quad (6)$$

on the supported vanadia catalysts are not large and do not significantly vary among the catalysts (minor differences may be lost in the experimental error). This suggests that the lower Arrhenius rate constant for methanol oxidation over silica supported vanadia catalysts is primarily due to a lower pre-exponential factor (an entropic effect). Weber recently proposed a model that predicted that differences in methanol oxidation TOFs among supported vanadia catalysts should be related to differences in an entropy effect in the Arrhenius constant, and the current findings between V_2O_5/SiO_2 and the other supported vanadia catalysts are consistent with this model.[43] However, this model does not account for the different methanol oxidation TOFs among the other supported vanadia catalysts because these differences are related to the equilibrium adsorption constant. Stiegman recently proposed that the energy level of the bridging V-O-Support bond should determine the reactivity of this bond, reflected in the Arrhenius rate constant, during methanol oxidation and the current findings between V_2O_5/SiO_2 and the other supported vanadia catalysts are also consistent with this model.[44] However, it appears that the differences in reactivity of the bridging V-O-S bond are mostly related to the adsorption step, K, rather than the Arrhenius rate constant, k, for the decomposition of surface methoxy to form formaldehyde. Thus, theoretical models of oxidation reactions should not only consider the surface reaction rate constants, k, but also the equilibrium adsorption constants, K, which may also significantly contribute to the overall specific reaction rates or TOF.

5 Influence of the Oxide Support upon Different Oxidation Reactions

The above studies with methanol oxidation over the different supported vanadia catalysts demonstrated the dramatic effect of the specific oxide support on the reaction TOF, approximately four orders of magnitude for methanol oxidation. Several other oxidation reactions over the same series of supported vanadia catalysts were also investigated to determine if the effect of the oxide support is a general phenomenon, and the results are listed in Table 2.

For all the oxidation reactions listed in Table 2, it was found that the specific oxide support had a significant effect on the oxidation reaction TOF. The methanol oxidation TOF was the most sensitive to the oxide support ($\sim 10^4$ factor), followed by butane oxidation [20,38] and selective catalytic reduction of NO by NH_3[45] ($\sim 10^2$ factor), and oxidation of CO[46] and SO_2[47] (~10 factor) were the least sensitive. The differences in order of magnitude of the support effect on the

Table 2 *Support Effect is a General phenomenon for Oxidation Reactions*

Oxidation reaction			*Support effect on TOF (for vanadia catalysts)*
CH_3OH	$\longrightarrow$	H_2CO+H_2O	$\sim10^4$
$CH_3CH_2CH_2CH_3$	$\longrightarrow$	maleic anhydride $+H_2O$	$\sim10^2$
$NO+NH_3$	$\longrightarrow$	N_2+H_2O	$\sim10^2$
CO	$\longrightarrow$	CO_2	~10
SO_2	$\longrightarrow$	SO_3	~10

• In general: $CeO_2 > ZrO_2 \sim TiO_2 > Nb_2O_5 > Al_2O_3 > SiO_2$

different reactions most likely reflect different reaction requirements of the various reactants. More detailed kinetic and mechanistic studies of these oxidation reactions, similar to the methanol oxidation studies reported above, are required to fully understand how the oxide support ligand affects these reactions. Furthermore, in general, the relative trend in TOF followed the pattern CeO_2> ZrO_2~TiO_2>Nb_2O_3>Al_2O_3>SiO_2 for these oxidation reactions. Thus, it appears that the bridging V-O-Support bond plays a critical role in all of these oxidation reactions and is a general phenomenon in oxidation reactions.

6 Number of Surface Sites Required for Oxidation Reactions

Information about the number of surface sites required for an oxidation reaction, or activation of the reactant molecule, can be obtained by examination of the variation of the TOF with surface vanadia coverage. In general, reactions requiring only one surface site will exhibit a TOF that is independent of the surface vanadia coverage (surface density of sites) and reactions requiring multiple surface sites will exhibit a TOF that increases with the surface vanadia coverage (surface density of sites). From such an analysis, the number of surface vanadia sites required for various oxidation reactions is presented in Table 3.

Table 3 *Number of Surface Sites required for Various Oxidation Reactions*

Oxidation reactions			*No. of required surface sites*
CH_3OH	$\longrightarrow$	H_2CO+H_2O	1
$CH_3CH_2CH_3$	$\longrightarrow$	$CH_3CHCH_2+H_2O$	1
$CH_3CH_2CH_2CH_3$	$\longrightarrow$	maleic anhydride $+H_2O$	1-2*
$NO+NH_3$	$\longrightarrow$	N_2+H_2O	1-2*
CH_4	$\longrightarrow$	H_2CO+H_2O	1
CO		CO_2	1
SO_2		SO_3	1

* Reactions more efficient over two adjacent sites

The oxidation of methanol to formaldehyde,[29] propane to propylene,[48] methane to formaldehyde,[49] CO to CO_2[46] and SO_2 to SO_3[47] require only one surface vanadia site. In the case of methanol oxidation, where the molecular mechanism is known, IR studies have confirmed that the surface methoxy intermediate is coordinated to only one surface vanadia species.[40] Such detailed molecular information is currently not available about the other oxidation reactions over the supported vanadia catalysts. The selective oxidation of butane to maleic anhydride[20,38] and the bimolecular NO/NH_3 reaction[30] exhibit a significant increase of the TOF with surface coverage and suggests that multiple sites, most probably two sites, are required for these oxidation reactions (see additional discussion below in Section 7). Although the oxidation of butane to maleic anhydride and the selective catalytic reduction of NO with NH_3 are more efficient over several adjacent vanadia sites, these reactions have also been shown to occur over isolated vanadia sites present at low surface vanadia coverages and on the silica support (see Section 2 above).[20,30,38] Consequently, all oxidation reactions can proceed over isolated surface vanadia sites, but oxidation reactions involving multisteps and bimolecular reactions proceed more efficiently over several adjacent surface vanadia sites.

7 Influence of Surface Metal Oxide Additives upon Oxidation Reactions

7.1 Promoters. – Many supported vanadia catalysts also possess secondary metal oxides additives that act as promoters (enhance the reaction rate or improve product selectivity). Some of the typical additives that are found in supported metal oxide catalysts are oxides of W, Nb, Si, P, etc. These secondary metal oxide additives are generally not redox sites and usually possess Lewis and Bronsted acidity.[50] Similar to the surface vanadia species, these promoters preferentially anchor to the oxide substrate, below monolayer coverage, to form two-dimensional surface metal oxide species. This is schematically shown in Figure 4.

Such secondary surface metal oxide additives, in general, only indirectly interact with the surface vanadia species via lateral interactions[50] or possibly even electronic effects through semiconducting oxide supports.[51] The lateral interactions may also slightly affect the ratio of polymerized to isolated surface vanadia species due to surface crowding by the secondary metal oxide additives.[50] The influence of the secondary metal oxide additives upon an oxidation reaction depends on the specific requirements of the reactions. For oxidation reactions requiring one surface vanadia site, such as the oxidation of methanol to formaldehyde, such secondary metal oxide additives have only a minor effect since they are generally not redox sites and the oxidation reaction only proceeds on the surface vanadia sites.[50] The acidic character of the secondary metal oxide additives, however, can sometimes lead to byproducts since the acid sites can catalyse the formation of dimethyl ether from methanol. For oxidation reactions most efficiently proceeding over adjacent surface redox-surface acid dual sites, such as the oxidation of butane to maleic anhydride[38,52]

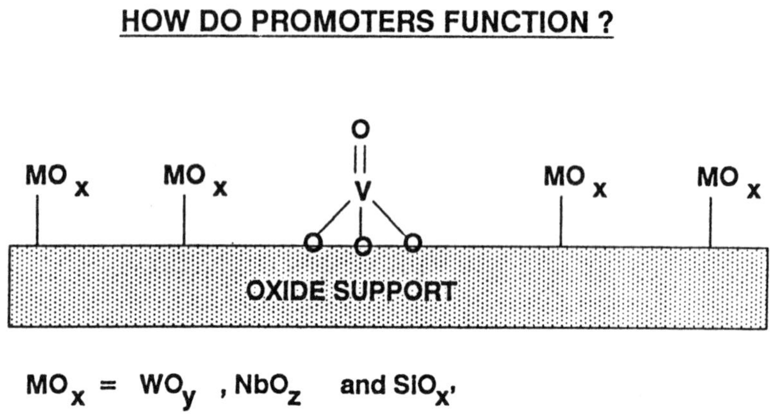

Figure 4 *How do promoters function?*

and the selective catalytic reduction of NO with NH_3,[30,53] significant enhancement in reaction rate and product selectivity by the presence of these acidic secondary metal oxide additives is observed. Thus, redox reactions requiring one surface vanadia site are generally not promoted by such secondary metal oxide additives, but redox-acid dual site reactions are generally promoted by such secondary metal oxide additives.

7.2 Poisons. – Unlike secondary surface metal oxide additives that indirectly interact with the surface vanadia sites via lateral interactions, poisons are surface metal oxide additives that directly interact with the surface vanadia sites and decrease the TOF. For example, the addition of surface potassium oxide to supported vanadia catalysts results in both a structural change and a reactivity change of the surface metal oxide species.[50] This interaction, at submonolayer coverages, reflects the attractive interaction between these two surface metal oxide species. The presence of the surface potassium oxide poison alters the V-O bond lengths and the ratio of polymeric and isolated surface vanadia species (favoring isolated surface vanadia species). The interaction of the surface potassium oxide poison with the surface vanadia species is schematically shown in Figure 5.

These structural changes are accompanied by significant reactivity modifications of the surface vanadia species. The addition of the surface potassium oxide species decreases the reducibility of the surface vanadia species in temperature programmed reduction (TPR) studies and the TOF for methanol oxidation.[23,50] The most likely reason for this behavior is that the surface potassium oxide species is intimately coordinated to the bridging V-O-Support bond and retards its participation in these redox processes. Thus, all oxidation reactions, involving one surface vanadia site as well as dual surface vanadia-acidic sites, will be retarded by the surface potassium oxide additive. The basic properties of the surface potassium oxide additive may also affect the product selectivity by

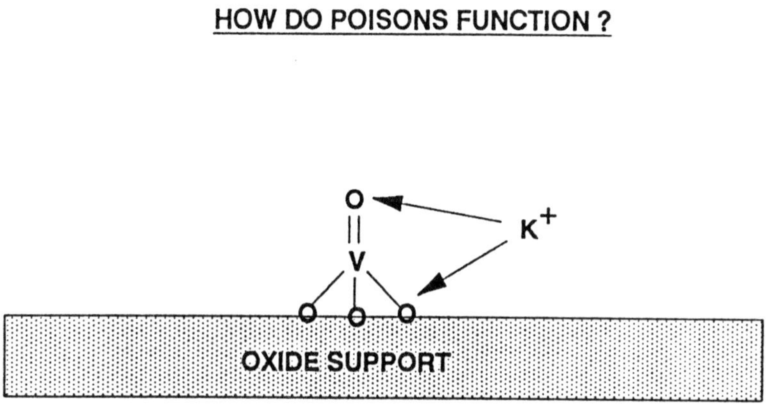

Figure 5 *How do poisons function?*

neutralizing surface Lewis and Bronsted sites. This may be a desirable feature for redox reactions where the acidic sites lead to unwanted byproducts (e.g., conversion of CH_3OH to CH_3OCH_3). Thus, the addition of basic poisons to supported vanadia catalysts retards the TOF for all oxidation reactions, but may enhance the product selectivity by neutralizing surface acid sites that may be responsible for side reactions.

8 Influence of the Specific Oxide Support Phase

The influence of the specific oxide support phase upon the structure and reactivity of the surface vanadia species was also recently investigated.[54] A series of titania-supported vanadia catalysts were synthesized over a series of TiO_2 supports possessing different phases (anatase, rutile, brookite and B). Raman and solid state vanadium-51 characterization studies revealed that the same surface vanadia species were present in all the different V_2O_5/TiO_2 catalysts[54]. The reactivity of the surface vanadia species on the different oxide supports was probed by methanol oxidation and the TOFs are shown in Figure 6 (all the catalysts contained 1% V_2O_5)

Essentially the same methanol oxidation TOFs were obtained on the different oxide supports. The Degussa P-25 titania support (90% anatase; 10% rutile) was also examined, as shown in Figure 6, because it possesses very low levels of surface impurities and represents a good reference sample. The invariance of the methanol oxidation TOF with the specific phase of the titania support reveals that the oxidation reaction is controlled by a local phenomenon, the bridging V-O-Support bond, rather than long range effects, the structure of the TiO_2 support. Thus, the phase of the oxide support does not appear to influence the molecular structure or reactivity of the surface vanadia species.

SUPPORT PHASE DOES NOT AFFECT OXIDATION REACTIONS

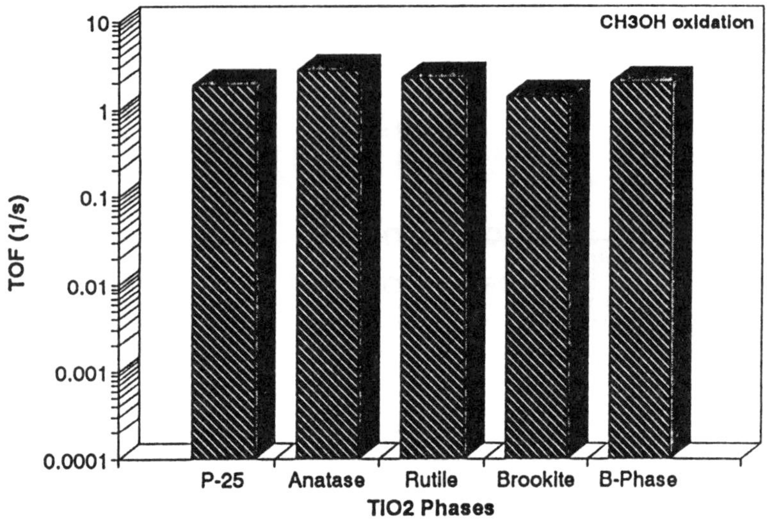

- **oxidation reaction controlled by stability of bridging V-O-Ti bond (a local phenomenon !)**

Figure 6 *Support phase does not affect oxidation reactions.*

9 Influence of Preparation Method

The influence of different preparation methods upon the structure and reactivity of supported vanadia catalysts has recently been investigated by several laboratories.[55-58] These studies concluded that the same surface vanadia species forms on a specific oxide support independent of the preparation method, even when starting with V_2O_5 crystals. The reason for this phenomenon is that vanadium oxide spontaneously disperses on oxide supports to form surface vanadia species, which is schematically depicted in Figure 7.

This spontaneous dispersion, or self-assembly, of vanadium oxide on oxide supports is driven by the minimization of the surface free energy of the system: the surface free energy of V_2O_5 is significantly less than the surface free energy of most typical oxide supports (e.g., Al_2O_3, TiO_2, ZrO_2, etc.).[58] Furthermore, the surface vanadia species are able to readily rearrange and surface diffuse at typical catalytic reaction temperatures because of the low melting point (695 °C) and, consequently, Tamman temperature (370 °C) of V_2O_5. Thus, the preparation method cannot influence the final state of equilibrated supported vanadia catalysts.

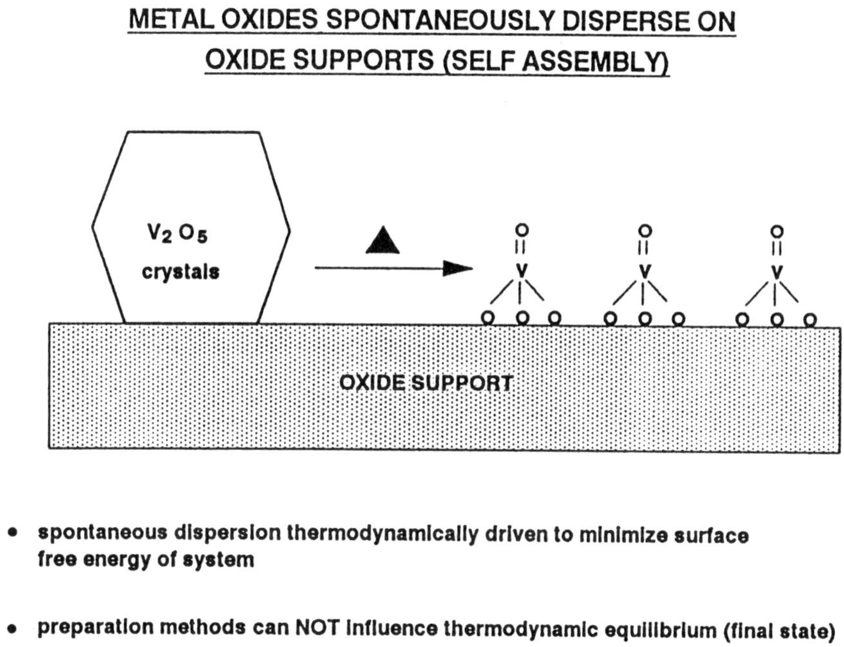

Figure 7 *Metal oxides spontaneously disperse on oxide supports (self assembly).*

10 Conclusions

The above molecular structural and reactivity studies provide the foundation for the molecular ingredients required to molecularly engineer supported vanadia catalysts for oxidation reactions. The number of active surface sites is determined by the surface vanadia coverage in the two-dimensional vanadia overlayer (a catalyst synthesis parameter). The specific activity of the active surface sites is controlled by the specific oxide support ligand (a catalyst synthesis parameter), but is not influenced by the specific phase of the oxide support. The specific activity of the active surface sites is not a function of surface vanadia coverage for oxidation reactions efficiently proceeding over isolated surface vanadia sites (oxidation of methanol to formaldehyde, methane to formaldehyde, propane to propylene, CO to CO_2, SO_2 to SO_3, etc.), but the specific activity is a strong function of surface vanadia coverage for oxidation reactions that proceed more efficiently over several adjacent surface vanadia sites (selective oxidation of butane to maleic anhydride, selective catalytic reduction of NO_x with NH_3 to N_2 and H_2O, etc.). The specific activity of the active surface sites can be further enhanced by the introduction of acidic secondary surface metal oxide additives (a catalyst synthesis parameter), but only for oxidation reactions that proceed more efficiently over several adjacent surface vanadia sites. The specific preparation method does not influence the nature of the active surface sites since the same surface vanadia sites are formed in equilibrated catalysts from all the different catalyst preparations. Thus, the critical molecular engineering or catalyst synthesis parameters that control the overall activity of supported vanadia catalysts during oxidation reactions are (1) the specific oxide support material (in general, $CeO_2 > ZrO_2 \sim TiO_2 > Nb_2O_5 > Al_2O_3 > SiO_2$), (2) the surface area of the oxide support (more of the two-dimensional vanadia overlayer can be accommodated with higher surface area supports, but the thermal stability of such supported vanadia catalysts decreases with increasing surface area), (3) the vanadia content of the final catalyst (monolayer surface coverage corresponds to ~1 wt.% V_2O_5 per 10 m^2/g of support), (4) the introduction of secondary acidic surface metal oxide additives (e.g., W, Nb, P, etc.), but only for oxidation reactions that proceed more efficiently over several adjacent sites. These molecular engineering principles have also been successfully extended to the synthesis of multilayered supported vanadia catalysts (e.g., a surface vanadia overlayer on a two-dimensional surface titania phase coordinated to a silica support[59]) as well as to other supported metal oxide oxidation catalysts (e.g., Mo[60], Cr[61,62] and Re[63]).

Acknowledgments

The financial support of the Division of Basic Energy Sciences, Department of Energy (grant no. DEFG02-93ER14350) and the National Science Foundation (grant no. CTS-9417981) is gratefully acknowledged. This paper is dedicated to my former and current students at Lehigh University who have contributed to advancing the catalysis science of supported metal oxide catalysts (F.D. Hardcastle, J.-M. Jehng, G. Deo, M. A. Vuurman, D. S. Kim, A. M. Turek, J. Datka, H. Hu, M. A. Banares, M. Osrtomecki, N. Arora, M. Kellner, Y. Cai, C.-B. Wang, J. Dunn, L. Burcham, X. Gao and P. R. Koppula).

References

1. C. L. Thomas, 'Catalytic Processes and Proven Catalysts', Academic Press, New York , 1970.
2. K. Weissermel and H.-J. Arpe, "Industrial Organic Chemistry", Verlag Chemie, Weinheim/New York, 1978.
3. J. Armor, 'Environmental Catalysis', ACS Symp. Series 552, Am. Chem. Soc., Washington, D.C., 1994.
4. G. Centi and P. Forzatti, 'Environmental Catalysis', SCI Publications, Rome, 1995.
5. B. Grzybowska and F. Trifiro, special issue of *Appl. Catal.* (in press).
6. G. C. Bond and J. C. Vedrine, *Catal. Today*, 1994, **20**, 1.
7. G. Centi, *Appl. Catal.* A, 1996, **147**, 267.
8. H. Bosch and F. J. J. G. Janssen, *Catal. Today*,1988, **2**, 369.
9. N. D. Spencer and C. J. Pereira, *J. Catal.*, 1989, **116**, 399.
10. N. Ohlms, *Catal. Today*, 1993, **16**, 247.
11. S. T. Oyama and G. A. Somorjai, *J. Phys. Chem.*, 1990, **94**, 5022.

12. R. H. H. Smits, K. Seshan, H. Leemreize and J. H. Ross, *Catal. Today*, **16**, 513.
13. B. Grzybowska, P. Mekss, R. Grabowski, K. Wcislo, Y. Barbaux and L. Gengembre, *Stud. Surf. Sci. Catal.*, 1994, **82**, 151.
14. J. G. Eon, R. Olier and J. C. Volta, *J. Catal.*, 1994, **145**, 318.
15. L. Owens and H. H. Kung, *J. Catal.*, 1993, **144**, 202.
16. P. J. Anderson and H. H. Kung, *Stud. Surf. Sci. Catal.*, 1993, **75**, 206.
17. A. Ramstetter and M. Baerns, *J. Catal.*, 1988, **109**, 303.
18. R. A. Overbeek, PhD Thesis, Technical University of Utrecht, The Netherlands, 1994.
19. W. Harding, K. E. Birkeland and H. H. Kung, *Catal. Lett.*, 1994, **28**, 1.
20. I. E. Wachs, J.-M. Jehng, G. Deo, B. M. Weckhuysen, V. V. Guliants and J. B. Benziger, *Catal. Today*, 1996, **32**, 47.
21. P. M. Michalakos, K. E. Birkeland and H. H. Kung, *J. Catal.*, 1996, **158**, 349.
22. P. Gellings, in 'Catalysis', ed. G. C. Bond and G. Webb, Specialist Periodical Reports, The Royal Society of Chemistry, London, UK, 1985, vol. **7**, 105.
23. G. C. Bond and S. F. Tahir, *Appl. Catal.*, 1991, **71**, 1.
24. G. Deo, I. E. Wachs and J. Haber, *Crit. Rev. Surf. Chem.*, 1994, **4**, 141.
25. I. E. Wachs and K. Segawa, in 'Characterization of Catalytic Materials', ed. I. E. Wachs, Butterworth-Heinemann, Stoneham, 1992, chapter 4.
26. I. E. Wachs, *Catal. Today*, 1996, **27**, 437.
27. G. T. Went, L.-J. Leu, R. R. Rosin and A. T. Bell, *J. Catal.*, 1992, **134**, 492.
28. S. T. Oyama, G. T. Went, K. T. Lewis, A. T. Bell, and G. A. Somorjai, *J. Phys. Chem.*, 1989, **93**, 6786.
29. G. Deo and I. E. Wachs, *J. Catal.*, 1994, **146**, 323.
30. I. E. Wachs, G. Deo, B. M. Weckhuysen, A. Andreini, M. A. Vuurman, M. De Boer and M. Amiridis, *J. Catal.*, 1996, **161**, 211.
31. S. Yoshida, T. Tanaka, T. Hanada, T. Hiraiwa and H. Kanai, *Catal. Lett.*, 1992, **12**, 277.
32. H. Eckert and I. E. Wachs, *J. Phys. Chem.*, 1989, **93**, 6796.
33. N. Das, H. Eckert, H. Hu, I. E. Wachs, J. F. Walzer and F. Feher, *J. Phys. Chem.*, 1993, **97**, 8240.
34. G. T. Went, S. T. Oyama and A. T. Bell, *J. Phys. Chem.*, 1990, **94**, 4240.
35. M. A. Vuurman, I. E. Wachs and A. M. Hirt, *J. Phys. Chem.*, 1991, **95**, 9928.
36. M. A. Vuurman and I. E. Wachs, *J. Phys. Chem.*, 1992, **96**, 5008.
37. G. Busca, Mater. *Chem. Phys.*, 1988, **19**, 157.
38. I. E. Wachs, J.-M. Jehng, G. Deo, B. Weckhuysen, V. V. Guliants, J. B. Benziger and S. Sundaresan, *J. Catal.* (in press).
39. U. Chowdhry, A. Ferretti, L. E. Firment, C. J. Machiels, F. Ohuchi, A. W. Sleight and R. H. Staley, *Appl. Surf. Sci.*, 1984, **19**, 360.
40. G. Busca, A. S. Elmi and P. Forzatti, *J. Phys. Chem.*, 1987, **91**, 5263.
41. I. E. Wachs, G. Deo, M. Juskelis and B. M. Weckhuysen, in 'Proc. Int. Symp. on Dynamics of Surfaces and Reaction Kinetics in Heterogeneous Catalysis', ed. G. F. Froment and K. C. Waugh, Elsevier, Amsterdam (in press).
42. G. Deo, L. Burcham and I. E. Wachs, to be published.
43. R. S. Weber, *J. Phys. Chem.*, 1994, **98**, 2999.
44. K. Tran, M. A. Hanning-Lee, A. Biswas, A. E. Stiegman and G. W. Scott, *J. Am. Chem. Soc.*, 1995, **117**, 2618.
45. M . D. Amiridis, R. Duval, G. Deo and I. E. Wachs, to be published.
46. J. Haber, G. Deo and I. E. Wachs, to be published.
47. J. Dunn, J.-M. Jehng, D. S. Kim, H. G. Stenger and I. E. Wachs, 'Preprints 11th Intern. Congr. Catal.', 1996, Po. 215.
48. T. C. Watling, G. Deo, K. Seshan, I. E. Wachs and J. A. Lercher, *Catal. Today*, 1996, **28**, 139.
49. Q. Sun, J.-M. Jehng, H. Hu, R. G. Herman, I. E. Wachs and K. K. Klier, *J. Catal.*, 1997, **165**, 91.
50. G. Deo and I. E. Wachs, *J. Catal.*, 1994, **146**, 335.

51. G. Allemany, N. Ferlazzo, P. Forzatti, L.Lietti, G. Busca, E. Giamello and F. Bregani, *J. Catal.*, 1995, **155**, 117.
52. F. Cavani and F. Trifiro, *Chemtech*, 1994, **24**, 18.
53. N.-Y. Topsoe, J. A. Dumesic, and H. Topsoe, *J. Catal.*, 1995, **151**, 241.
54. G. Deo, A. M. Turek, I. E. Wachs, T. Machej, J. Haber, N. Das, H. Eckert and A. M. Hirt, *Appl. Catal.* A., 1992, **91**, 27.
55. T. Machej, J. Haber, A. M. Turek and I. E. Wachs, *Appl. Catal.*, 1991, **70**, 115.
56. G. Centi, D. Pinelli, F. Trifiro, G. Ghoussoub, M. Guelton and L. Gengembre, *J. Catal.*, 1991, **130**, 238.
57. J. Haber, T. Machej, E. M. Serwicka and I. E. Wachs, *Catal. Lett.*, 1995, **32**, 101.
58. H. Knozinger and E. Taugler in 'Catalysis', ed. J. J. Spivey, Specialist Periodical Report, The Royal Society of Chemistry, Cambridge, UK, 1993, vol. **10**, 1.
59. J.-M. Jehng and I. E. Wachs, *Catal. Lett.*, 1992, **13**, 9.
60. H. Hu and I. E. Wachs, *J. Phys. Chem.*, 1995, **99**, 10911.
61. D. S. Kim and I. E. Wachs, *J. Catal.*, 1993, **142**, 166.
62. J.-M. Jehng, I. E. Wachs, B. M. Weckhuysen and R. A. Schoonheydt, *J. Chem. Soc., Faraday Trans.*, 1995, **91**, 953.
63. D. S. Kim and I. E. Wachs, *J. Catal.*, 1993, **142**, 166.

3

Zeolite-catalysed Alkylation of Polynuclear Aromatics

BY YOSHIHIRO SUGI AND YOSHIHIRO KUBOTA

1 Introduction

Shape-selective reactions occur by differentiating reactants, products, and/or reaction intermediates according to their shape and size in sterically restricted environments of the pore structures of microporous crystals[1-6]. If all of the catalytic sites are located inside a pore that is small enough to accommodate both the reactants and products, the fate of the reactant and the probability of forming the product are determined by molecular size and configuration of the pore as well as by the characteristics of its catalytic center, *i.e.*, only a reactant molecule whose dimension is less than a critical size can enter into the pore and react at the catalytic site. Furthermore, only product molecule that can diffuse out through the pore will appear in the product.

Zeolites are the most promising microporous materials for achieving highly shape-selective catalysis because their pores are uniformly distributed and have dimensions allowing both the organic reactants and products to enter, to react, and to leave.[1]

The catalytic alkylation of aromatics using zeolites has been the subject of much research[1-12] because it is essential to match the dimensions between reactants, products, and zeolite pores in order to achieve highly shape-selective catalysis. The pores of ten-membered ring zeolites (*e.g.*, HZSM-5) are especially suitable for highly shape-selective methylation of mononuclear aromatics, while twelve-membered ring zeolites, such as H-mordenite (HM), HY, *etc.*, have pore sizes that are too large to control the selectivity for the reaction.[8-11] On the other hand, HM is useful for catalysing the shape-selective alkylation and other related reactions of mononuclear aromatics such as ethylbenzene using bulky alkylating agents.[12]

Recently, the synthesis of symmetrically substituted dialkylpolynuclear aromatic hydrocarbons, such as 2,6-diisopropylnaphthalene and 4,4′-diisopropylbiphenyl has been studied because they are superior candidates of components for advanced materials.[4,7] Polynuclear aromatics require larger space for the transition state intermediate composed of reactants and acid sites inside the pores than do mononuclear reactants. For these reasons, twelve-membered ring zeolites, especially HM, are suitable for the formation of the smallest products although the selectivity varies with reactants and zeolites. In this paper, we review the shape-selective alkylation of polynuclear aromatics catalysed by zeolites.

2 Mechanisms of Shape-selective Catalysis

Three types of shape-selective catalysis are shown in Figure 1. They differ depending on whether the pore size limits the entrance of reactant molecules, the departure of product molecules, or the formation of certain transition states[1,13]:

(1) *Reactant selectivity* occurs when some of the molecules in a reaction mixture can enter the pores and react in the catalyst pores. However, the molecules that are too large to diffuse through the pores cannot react.

(2) *Product selectivity* occurs when some of the products formed in the catalyst pore are too bulky to diffuse out, being converted to less bulky molecules (*e.g.*, by equilibration or cracking). The large product molecules, which cannot diffuse out, may eventually deactivate the catalytic sites by blocking the pores.

(3) *Restricted transition-state selectivity* occurs when certain reactions are pre-

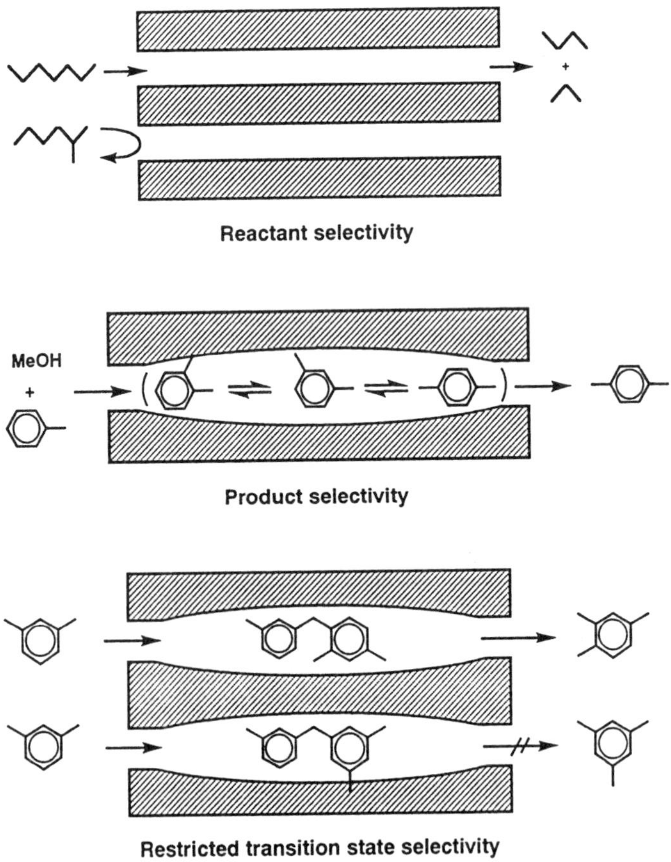

Figure 1 *Mechanisms of shape-selective catalysis.*

vented because the corresponding transition state would require more space than available inside the pores. Reactions requiring smaller transition states proceed unhindered to form smaller product molecules.

Although a clear distinction among these mechanisms is difficult, there is an important difference between *product selectivity* and *restricted transition-state selectivity mechanisms*. In the former mechanism, the product composition inside the pores should either be close to equilibrium, or the selectivity for the products inside the pores should be lower than that for bulk products. However, the selectivity for the narrowest isomer of the encapsulated products should be as high as that of bulk products in the latter mechanism.

Derouane and his co-workers[14,15] proposed that "hilly" environments offered by pore openings, cut channels, and/or cavities at the external surface of zeolites will preferentially adsorb and shape reactant molecules depending on their stereochemistry and their ability to optimize their van der Waals interaction with the framework, *i.e.* their capacity to "nest". Adsorption will be favored for molecules (or intermediates) which can easily adapt their geometry.

Derouane and Gabelica[16] proposed *molecular traffic control* as another type of shape-selectivity that could occur in zeolites having more than one type of intersecting pore system. Here, reactant molecules may preferentially enter into the catalyst through one pore system while the products diffuse out through the other, thereby minimizing counter diffusion and increasing the reaction rate.

Zeolite catalysed alkylation of polynuclear aromatics is considered to be simultaneously governed by several mechanisms. To achieve highly shape-selective catalysis, it is essential that the pore size precisely corresponds to the molecular dimensions of reactants and products, and to the transition state of the reaction intermediates.

3 Alkylation of Biphenyl*

3.1 Isopropylation

3.1.1 Catalysis over Typical Zeolites – Table 1 summarizes results of the catalysis with typical zeolites in liquid phase isopropylation of biphenyl.[17-19] The isopropylation over HY(5.8)† and HL(6.1) zeolites as well as amorphous silica-alumina (SA(4.3)) was non-regioselective in the formation of three isopropylbiphenyl (IPBP) isomers. The selectivity for 4-IPBP among IPBP isomers over HY at 200 °C was as low as 41%, whereas selectivities for 2- and 3-IPBP were as high as 36% and 23%, respectively. Diisopropylbiphenyl (DIPB) isomers were also obtained non-regioselectively, with the selectivity for 4,4′-DIPB being less than 11% in DIPB isomers. However, the selectivity of thermodynamically more stable isomers, 3-IPBP among IPBP isomers, and 3,4′- and 3,3′-DIPB among DIPB isomers increased at 250 °C. The catalytic feature of HL and SA also resembled that of HY. The selectivity for 4-IPBP over HM(23) at 250 °C was more than

* The selectivity of an isomer is shown by the percentage in all isomers for each alkylate unless otherwise stated.

† The number in the parenthesis expresses the SiO_2/Al_2O_3 ratio of the zeolite.

70%, whereas it was less than 5% for 2-IPBP. For the DIPB isomers, the narrowest isomer, 4,4′-DIPB, was also formed in high selectivity (*ca.* 80 %). HZSM-5(50) showed low activity even at 300 °C.

Table 1 *Isopropylation of biphenyl catalysed by typical zeolites*[a]

Catalyst	*Reaction*		*Product composition (%)*		*Selectivity of IPBP (%)*			*Selectivity of DIBP (%)*		
(SiO_2/Al_2O_3)	*temp. (°C)*	*Conv. (%)*	*IPBP*	*DIBP*	*2-*	*3-*	*4-*	*4,4′-*	*3,4′-*	*3,3′-*
HM(23)	180	16	89	11	7	20	74	75	16	2
	250	48	73	27	5	24	71	78	14	2
HY(5.8)	200	76	60	40	36	23	41	5	8	7
	250	83	61	33	7	48	45	11	22	13
HL(6.1)	200	82	54	36	39	18	43	10	8	6
	250	84	53	47	29	25	46	10	13	6
SA(4.3)	180	67	62	38	36	15	49	16	9	5
	250	84	48	39	18	32	50	25	26	8
HZSM-5(50)	300	6	100	0	16	30	54	–	–	–

[a] Reaction conditions: biphenyl, 50 mmol; propylene, 100 mmol; solvent, *trans*-decalin, 20 ml; catalyst, 1 g; period, 4 h.

3.1.2 Isopropylation over H-Mordenite – Figure 2 shows the changes in the selectivity for IPBP and DIPB isomers over HM(220) at 250 °C under propylene pressure of 0.8 MPa.[17-29] Highly shape-selective isopropylation occurred to yield predominantly the least bulky 4-IPBP among IPBP isomers, and 4,4′-DIPB among DIPB isomers. 4,4′-DIPB was yielded with the consumption of 4-IPBP with an accumulation of 2- and 3-IPBP isomers, and the selectivity for 4,4′-DIPB was almost constant during the reaction. The highly selective formation of 4,4′-DIPB shows that the isopropylation proceeds by a consecutive mechanism, *i.e.*, biphenyl is isopropylated to form predominantly 4-IPBP, and the isopropylation of 4-IPBP yields 4,4′-DIPB regioselectively in the second stage.

H-mordenite has two types of pores which are perpendicular to each other: twelve-membered ring elliptical pores (0.67×0.71 nm) and eight-membered ring pores (0.29×0.57 nm).[30] The active centers in the catalyst should be inside the elliptical pores because biphenyl cannot enter into the latter pores; thus, these pores restrict the transition state of the isopropylation of biphenyl to form the least bulky products: 4-IPBP among IPBP isomers, and 4,4′-DIPB among DIPB isomers. However, the formation of DIPB isomers with 2- and 3-isopropyl groups was prevented in the pores because the corresponding transition states have bulky conformations that require larger space than are available at the acid sites.

3.1.3 Effect of the Dealumination of H-Mordenite – The acid catalysis depends on the number and strength of acid sites on the catalyst. However, the catalytic

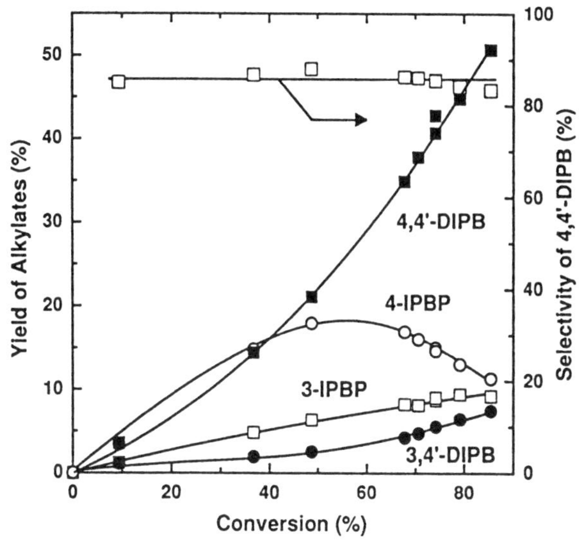

Figure 2 *The isopropylation of biphenyl over HM(220). Reaction conditions: biphenyl, 400 mmol; HM(220), 2 g; propylene pressure, 0.8 MPa; temperature, 250°C.*

activities are not always proportional to the acid density of zeolites. Coke-deposition inside the pores is one of the major reasons for the deactivation.[31-35] In the large-pore zeolites, such as HM and HY, the coke deposited in the pores is usually aromatic hydrocarbons having several nuclei. These deposits prevent acid catalysis because they choke the catalytically active sites in the pores. Most of the dense acid sites do not contribute to acid catalysis but promote coke-deposition. For these reasons, the dealumination of the zeolites is one of the effective methods for reduction of coke-deposition and for enhancement of catalyst performance even though the acid density decreases.[36-38]

Figure 3 shows the effect of the dealumination of HM in the isopropylation of biphenyl under propylene pressure of 0.8 MPa.[21,22] The change of the activity depending on SiO_2/Al_2O_3 ratio corresponds well to the change of the selectivity for the products. The activity and the selectivity for 4,4′-DIPB were low over HM with the low SiO_2/Al_2O_3 ratio although this catalyst has highly dense acid sites. However, yield of 4,4′-DIPB was enhanced by the dealumination, and, thus, HM(220) exhibited the highest yield despite having fewer acid sites.

Figure 4 shows TG profiles of the catalysts used in the isopropylation.[20,21,27] The amount of coke observed at around 600 °C decreased with the dealumination of HM. Volatile organic compounds, which are ascribed to isopropylated biphenyls encapsulated inside the pores, were also found at 300-350 °C for dealuminated HM with the SiO_2/Al_2O_3 ratio higher than 70 (*see below*). These results suggest that most of the volatile compounds were converted to coke

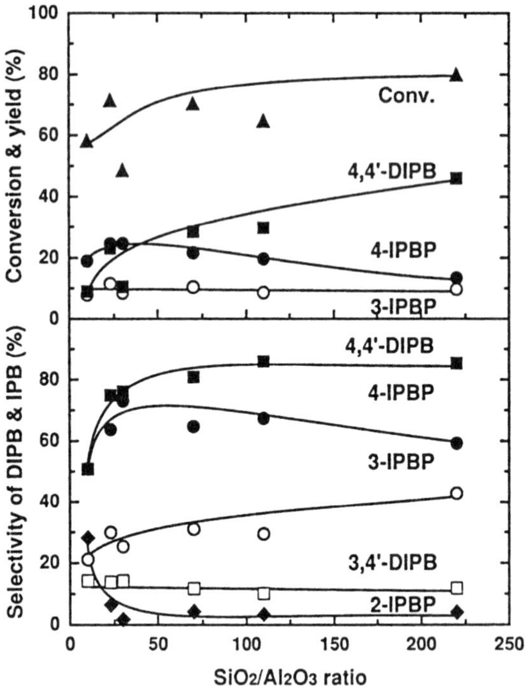

Figure 3 *Effect of the dealumination of HM on the isopropylation of biphenyl. Reaction conditions: biphenyl, 200 mmol; HM, 2 g; propylene pressure, 0.8 MPa; temperature, 250°C; period, 4h.*

deposits in the pores for HM with the low SiO_2/Al_2O_3 ratio, and that most of the acid sites in the pores are not catalytically active. A high silica HM such as HM(220) effectively catalyses shape-selective alkylation with the minimum coke deposition.

Coke deposits were formed from biphenyl over dealuminated HM only in the presence of propylene, although the amount of coke was less.[26,27] Coke deposition occurred in a short period after starting the reaction, and by the contact of 4,4′-DIPB with HM even in the absence of propylene.[26] These results suggest that the isopropylated biphenyls produce coke by dehydrogenative condensation at their isopropyl groups on acid sites. Propylene oligomers were formed during the reaction.[26] They are alternative precursors of deposited coke.

The effect of the dealumination of HM on encapsulated products inside the catalysts used for the reaction is shown in Figure 5.[28,29] The selectivity for 4,4′-DIPB inside the pores was almost constant over all HM zeolites although the selectivity for 4,4′-DIPB in bulk reaction products varied with the SiO_2/Al_2O_3 ratio. These results suggest that the shape-selective isopropylation occurs inside the pores even over HM with the low ratio. The low catalytic activity and the low

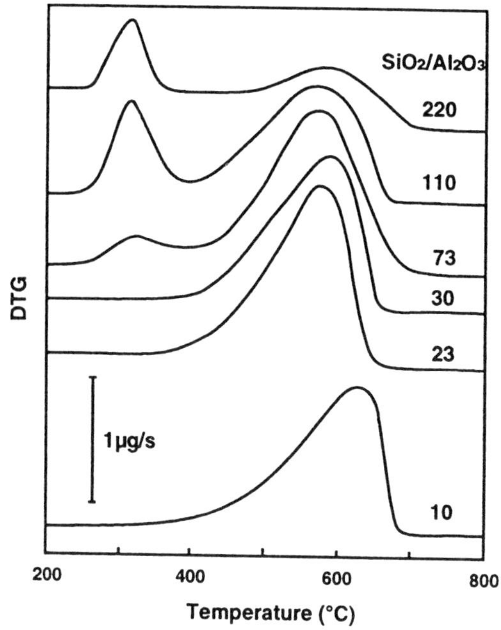

Figure 4 *Effect of the dealumination of HM on coke deposition in the isopropylation of biphenyl. Reaction conditions were the same as in Figure 3.*

selectivity for 4,4′-DIPB in bulk reaction products over HM with the low SiO_2/Al_2O_3 ratios were due to the non-regioselective catalysis at the external acid sites, which were still active after the pores were blocked. The high selectivity for 4,4′-DIPB in encapsulated DIPB isomers indicates that a *restricted transition-state mechanism* is in effect over HM, while a *product selectivity mechanism* is not.

Lee and his co-workers[39-41] reported that highly dealuminated mordenites were exceptionally active for the isopropylation of biphenyl at 250 °C, and the yield of 4,4′-DIPB increased with increasing the SiO_2/Al_2O_3 ratio as shown in Table 2. After 20 h, DHM-2X with SiO_2/Al_2O_3 ratio of 2600 prepared from Zeolon 100 gave an extremely high conversion of biphenyl (98%) and a high yield of 4,4′-DIPB (72%). These phenomena occur because the dealumination not only reduces the number of acid sites but also modifies the pore distribution, resulting in an increase in the volume of mesopores with radii of 2~200 nm. If the alkylation is diffusion limited, then the high yield by DHM-2X may be due to the improved diffusion caused *via* the creation of mesopores. The high yield of 4,4′-DIPB could also be related to a lower deactivation rate in the DHM. From XRD and adsorption data, they proposed a biphenyl packing mechanism involving a partial overlap of the aromatic rings, being similar to the two-layer packing of benzene in the twelve-membered ring pores of mordenite as described

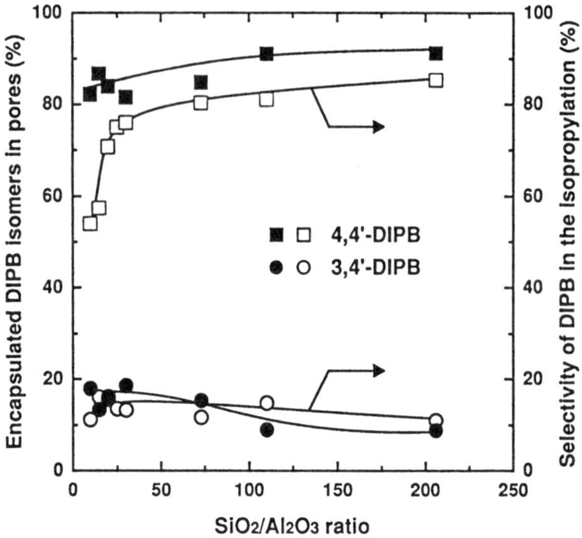

Figure 5 *Effect of the dealumination on the product distribution of encapsulated DIPB isomers inside pores and of bulk products in the isopropylation of biphenyl. Reaction conditions are the same as in Figure 3.*

by Itabashi and his co-workers.[42] Since the shape-selectivity appears to be related to both the packing of the reactants and the geometry of the mordenite pore, this phenomenon is called *reactant-assisted shape-selectivity.*[39] The occupation of the 12-membered ring pores in mordenite by polynuclear aromatics suggests that propylene may diffuse into the zeolite pores, at least partially, *via* the eight-membered ring pores. This high yield of 4,4'-DIPB is a clear example of *molecular traffic control.*[16,39]

Table 2 *Isopropylation of biphenyl over highly dealuminated H-mordenite*[a]

Catalyst	*SiO_2/Al_2O_3*	*Yield of 4,4'-DIBP (%)* 1 h	20 h	*Surface area (m^2/g)*	*Pore volume (cm^3/g)* micro	macro[b]
Zeolon 100	10	0	0.8	319	0.133	0.061
Zeolon 1X	144	3	28.6	465	0.184	0.087
Zeolon 2X	213	15	41.4	496	0.170	0.123
DHM-1X	256	35	56.2	428	0.208	0.200
DHM-2X	2600	56	72	382	0.149	0.227

[a] Reaction conditions: biphenyl, 500 g; catalyst, 10 g; propylene, 0.7 MPa; temperature, 250 °C; stirring rate, 2000 rpm.[b] pore radius, 2–200 nm.

3.1.4 Effect of Propylene Pressure on Selectivity – The partial pressure of propylene is also one of key factors for selective formation of 4,4′-DIPB.[22-25] The high partial pressure of propylene effectively enhanced the isopropylation, but the selectivity for 4,4′-DIPB decreased at lower partial pressures over HM(220), as shown in Figure 6.[24,25] However, the isomerization of 4-IPBP did not occur at any propylene pressures. Because 3,4′-DIPB is a more thermodynamically stable isomer than 4,4′-DIPB,[43] this decrease of the selectivity was ascribed to the isomerization of 4,4′-DIPB to 3,4′-DIPB, not to the lower selectivity to 4,4′-DIPB. Figure 7 shows the effect of propylene pressure on the selectivities for 4,4′-DIPB in bulk and encapsulated products. The selectivity of 4,4′-DIPB inside the pores was almost constant at every pressure. These results indicate that the isomerization does not occur inside the pores but at the external acid sites. The effect of the pressure on the isomerization of 4,4′-DIPB was similar to that on the isopropylation of biphenyl. 4,4′-DIPB itself isomerizes significantly to 3,3′- and 3,4′-DIPB over the catalyst in the absence of propylene. However, no significant isomerization of 4,4′-DIPB occurred in the presence of sufficient propylene pressure. On the other hand, the selectivity of 4,4′-DIPB in encapsulated DIPB isomers was almost constant at any pressure. These differences support the hypothesis that the isomerization of 4,4′-DIPB to 3,4′-DIPB occurs on the external surfaces. The isomerization of 4,4′-DIPB under high pressures is considered to be retarded by the preferential adsorption of propylene on acid

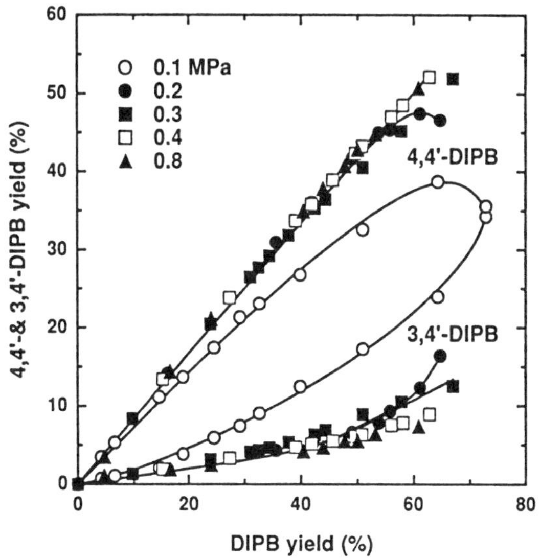

Figure 6 *Effect of propylene pressure on the yield of 4,4′- and 3,4′-DIPB in the isopropylation of biphenyl. Reaction conditions biphenyl, 400 mmol; HM(220), 2 g; propylene pressure, 0.1-0.8 MPa; temperature, 250 °C.*

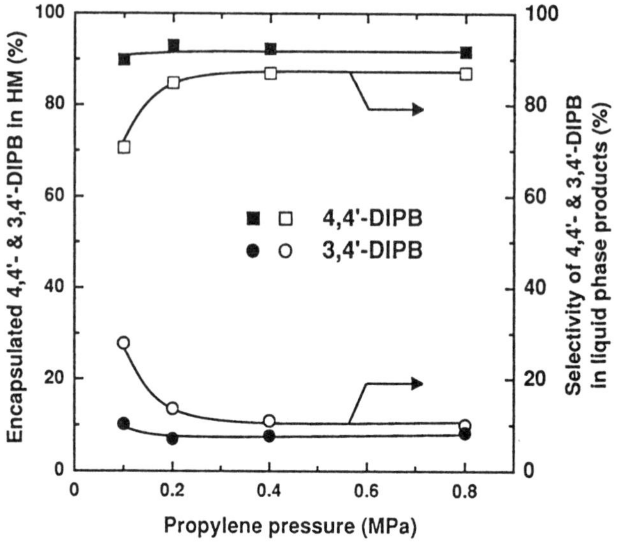

Figure 7 *Effect of propylene pressure on the product distribution of encapsulated 4,4′- and 3,4′-DIPB inside the pores and of the bulk products in the isopropylation of biphenyl. Reaction conditions: biphenyl, 200 mmol; HM(206), 1 g; propylene pressure, 0.1-0.8 MPa; temperature, 250 °C; period, 4 h.*

sites. However, the adsorption of 4,4′-DIPB should predominate over that of propylene under the low pressure, and, thus, the isomerization of 4,4′-DIPB occurs at external acid sites.

Matsuda and his co-workers also showed that extensive isomerization of 4,4′-DIPB on a HM(20) catalyst was observed at high conversion of biphenyl in the presence of limited amounts of propylene (initial propylene/biphenyl ratio=1.0).[44] However, the selectivity was almost constant during the reaction although the selectivity for 4,4′-DIPB was less than that of HM(10) catalyst. They suggest that the low isomerization activity of HM(10) is related to its weak acid strength.

Another interesting feature of the reaction of 4,4′-DIPB is that no further isopropylation to polyisopropylbiphenyls was observed under high pressure of propylene.[20-24] There is not enough space in the pores for the transition state of further isopropylation of 4,4′-DIPB. This is one of the reasons for the highly regioselective formation of 4,4′-DIPB.

Takahata and his co-workers found an increase in the selectivity for 4,4′-DIPB with increasing propylene pressure over HM with a low SiO_2/Al_2O_3 ratio.[45,46] Fellman suggested that the increase in selectivity was due to the accumulation of propylene at active sites.[7] However, it is obvious from above discussion that the change of the selectivity for 4,4′-DIPB with propylene pressure is due to the decrease of the isomerization of 4,4′-DIPB.

3.1.5 Effects of the Modification of H-Mordenite – Matsuda and his co-workers investigated the alkylation of biphenyl with propylene using mordenites (SiO_2/Al_2O_3=19) of different levels of proton exchange.[44] Mordenite, HNaM(19), of which 55% sodium cations were exchanged with protons, had the highest catalytic activity. Propylene was selectively consumed by isopropylation on mordenite at a proton exchange level below 55%. These results suggest that the strong acid sites promote side reactions of propylene. 4,4′-DIPB isomerized to 3,4′- and 3,3′-DIPB during the alkylation although the isomerization was suppressed by the reduction of acid strength. They concluded that the activity and the selectivity of mordenite zeolites are enhanced by the reduction of acid strength due to the suppression of undesirable side reactions.

Non-regioselective alkylation on external acid sites decreases the selectivity for 4,4′-DIPB. The deactivation of external acid sites is essential for high regio-selectivity. Matsuda and his co-workers[47,48] described that the modification of the external surface of HM with tributyl phosphonate (TBP) reduces the decrease in the 4,4′-DIPB yield during the reaction. The decrease in yield over unmodified HM is caused by the isomerization of 4,4′-DIPB at the external acid sites. The selectivity for 4,4′-DIPB was enhanced by the modification, and modified HM(10) catalyst showed the highest selectivity. Because TBP cannot enter into the pores, the modification deactivates acid sites only at external surfaces without changing the properties of acid sites inside the pores. The cracking activity of 1,3,5-triisopropylbenzene (1,3,5-TIPB) was greatly reduced by the modification although that of cumene was unchanged. These results indicate that a neutralizing of the external acid sites with TBP prevents the isomerization of 4,4′-DIPB. Matsuda and his co-workers examined the consumption of propylene during the alkylation.[44] Propylene was consumed by the alkylation of biphenyl over HM(10), while undesirable side reactions of propylene occurred extensively over HM(20). The consumption of propylene by the alkylation of biphenyl was enhanced by the modification with TBP. They conclude that the reduction of acid strength enhances the selectivity for propylene alkylation.

Sugi and his co-workers found that ceria modification of HM effectively deactivated the external acid sites in the isopropylation of biphenyl[49] and naphthalene,[50,51] as will be discussed in the section 4.2.3.

3.2 Isopropylation over HY, HL and Other Zeolites – The catalysis over HY and HL zeolites is quite different from that over HM. The product distribution over HY and HL resembled that of SA: the formation of 4,4′-DIPB was non-regioselective (Table 1).[18,19,43] These catalyses are not controlled by the environments of the pores because both zeolites have sufficiently large pores for the transition states form all the IPBP and DIPB isomers. Instead, the product distribution markedly changes by the reaction temperature. At low temperatures, the products are determined by the reactivity at 2- and 4-positions of biphenyl for electrophilic alkylation to yield principally 2- and 4-IPBP for IPBP isomers and

DIPB isomers with 2-isopropyl groups. This is due to the higher electron density of the 2- and 4-positions compared to the 3-position. However, the amounts of 3- and 4-IPBP, 3,4′- and 3,3′-DIPB increased with increasing temperature. These isomers may be formed by the isomerization of IPBP and DIPB isomers with 2-isopropyl groups because the former isomers are thermodynamically more stable isomers than the latter. These results show that the isopropylation of biphenyl over HY, HL, and SA is controlled by the reactivity of the reactant molecules at low temperatures, and by the thermodynamic stability of the product molecules at higher temperatures.

Matsuda and his co-workers examined the isopropylation of biphenyl over H-offretite and SAPO-11.[52] H-Offretite has pore size of 0.67×0.68 nm, but its pores have cages in the channels.[30] SAPO-11 has pores of 0.63×0.39 nm.[30] H-Offretite was less selective for the formation of 4,4′-DIPB than HM although their catalytic activities were comparable. SAPO-11 exhibited a comparable selectivity for 4,4′-DIPB to HM although catalytic activity was low. The formation of 4,4′-DIPB over these two zeolites was shape-selective compared to the thermodynamically attainable level.

Loktev and Chekriy showed that HZSM-12 produced 4,4′-DIPB with much higher selectivity than HM, H-beta (Hβ), HY, and HZSM-5 using flow type reactor at 200-270 °C.[53] HM yielded predominantly 3,4′-DIPB among DIPB isomers under the conditions. HY and HZSM-5 extensively catalysed propylene oligomerization.

3.3 Alkylation with Other Alkenes over H-Mordenite – The steric restriction of the transition state is essential for high selectivity in the isopropylation of biphenyl. The selectivity should be highly dependent on the size of the alkene.

The ethylation of biphenyl gave quite different results than the isopropylation.[54-56] The selectivities for ethylbiphenyl (EBP) isomers were nearly in the ratio 4-EBP : 3-EBP : 2-EBP = 1 : 2 : 2 at an early stage.[54,55] This means that the ethylation of BP to EBP isomers is controlled kinetically. However, these three EBP isomers act differently at higher conversion. The yield of 4-EBP reached maximum at 40-50% of the conversion but decreased with the further reaction. The yield of 3-EBP increased at early and middle stages and was saturated at the late stage. However, the formation of 2-EBP increased monotonically during the reaction. These results suggest that these three isomers have different reactivities for the further ethylation to give diethylbiphenyl (DEBP) isomers.[55] The narrowest isomer, 4-EBP, was consumed at the highest rates to yield predominantly DEBP isomers having 4-ethyl group, *i.e.* 2,4-, 3,4-, 3,4′-, 2,4′-, and 4,4′-DEBP. The formation of 3,3′-DEBP suggests the participation of 3-IPBP. The yield of 2,4-, 3,4-, 3,4′-, and 2,4′-DEBP increased with the conversion while the yield of 4,4′-DEBP was constant during the reaction. These results show that the steric restriction of the transition state composed of biphenyl, ethylene, and acid sites allows the shape-selective formation of the narrowest isomer, 4,4′-DEBP.

The alkenes higher than butenes gave the similar product patterns for alkylbiphenyls (4-ABP) and 4,4′-dialkylbiphenyls (4,4′-DABP) in the alkylation of biphenyl.[18] In particular, 2-methylpropene, which gives the most bulky carbonium ions, yielded selectively 4-*t*-butylbiphenyl and 4,4′-di-*t*-butylbiphenyl. With 1-hexene, a mixture of biphenyls with 2- and 3-hexyl groups (2- and 3-HBP) was produced in an approximate ratio of 3:1. 3-HBP was formed by the alkylation after the isomerization of hexene-1 to hexene-2 and had a higher *para*-selectivity than 2-HBP.

3.4 Isopropylation and Ethylation of 4-Alkylbiphenyls over H-Mordenite – The isopropylation and the ethylation of 4-alkylbiphenyls (4-ABP) (alkyl = methyl, ethyl, isopropyl) were examined to elucidate the interaction at the transition state in the pores.[55] 4-Methyl-4′-isopropylbiphenyl (4,4′-MIPB), 4-ethyl-4′-isopropylbiphenyl (4,4′-EIPB), and 4,4′-DIPB, were obtained selectively from the isopropylation of 4-ABP. The selectivity for 4,4′-DABP increased in the order: 4,4′-MIPB < 4,4′-EIPB < 4,4′-DIPB. The selectivity for 4-IPBP was *ca.* 70 % at the early stage of the isopropylation of biphenyl. Moreover, in the isopropylation of *p*-terphenyl (TP), the selectivity for 4-isopropyl-*p*-terphenyl was higher than 90% at the early stage of the reaction. These results show that the 4-substituent group enhanced the shape-selective catalysis inside the pores in the order: none < methyl < ethyl < isopropyl < phenyl, corresponding to the order of the reactant bulkiness. On the other hand, the selectivity for 4,4′-DABP in the ethylation was much lower than that in the isopropylation although bulkiness of the 4-substituent reflected the selectivity for the former reaction. The selectivity for 4,4′-EIPB in the isopropylation of 4-EBP was much lower than that in the ethylation of 4-IPBP. These results show that the bulkiness of the alkene affects the selectivity more than does the 4-substituent.

4 Alkylation of Naphthalene*

4.1 Methylation of Naphthalene over Zeolites – Fraenkel and his co-workers examined the vapor phase methylation of naphthalene with methanol over various zeolites.[57,58] HM and HY zeolites showed high conversions while HZSM-5 gave much lower conversion. The catalysis with HM and HY yielded a complex mixture containing a wide spectrum of products ranging from methylnaphthalenes (MN) to pentamethylnaphthalenes (PMN). The combined amount of 2,6- and 2,7-dimethylnaphthalene (DMN) isomers constituted only 35% and 32% of the DMN isomers for HM and HY, respectively, with their β/α ratios (2-/1-MN) being less than 2. This indicates that the pores of HM and HY do not contribute to the shape-selective methylation of naphthalene and 2-MN. HZSM-5 showed a high β-selectivity in the formation of MN and DMN isomers: MN had a much larger β/α ratio of 8.1, and 2,6- and 2,7-DMN isomers constituted 65 to 80% of the DMN isomers. Similar β-selectivity was observed in the methylation of 2-MN. These shape-selectivities can be explained by the size

* The selectivity of an isomer is shown by the percentage in all isomers for each alkylate unless otherwise stated.

differences between the relevant molecules and zeolite pores. The pores of HM and HY can accommodate all the DMN isomers and allow their rapid diffusion to occur during non-regioselective methylation. The authors proposed that naphthalene and 2-MN do not penetrate the ten-membered ring of HZSM-5, and that β-selective methylation occurs in the half-cavities of external surface sites, whose size matches that of the small-size naphthalenes.[58] Derouane and his co-workers subsequently coined the term "nest effect" to represent this phenomenon.[14,15]

Weitkamp and Neuber[59] found that high selectivities for the narrow methylation products, *i.e.*, 2,6-, 2,7-, and 2,3-DMN, were obtained over HZSM-5 and HZSM-11 in the methylation of 2-MN. Over these catalysts, the isomerization to 1-MN always decreased with time on stream, whereas the methylation of the DMN isomers either remained nearly constant or only slightly increased. After ~10 h, 50% of the DMN isomers were composed of 2,6-DMN, with all three of these narrow DMN products accounting for more than 90% of the product. These results clearly indicate that shape-selectivity becomes more pronounced with time on stream. Over an HZSM-5 catalyst precoked at 400 °C by converting methanol alone (into hydrocarbons), the initial yield of 1-MN was significantly reduced (to 1.5% *vs.* 4% for a fresh catalyst), with the initial content of 2,6- and 2,7-DMN in the DMN isomers being greater (84%) on the precoked catalyst than on the fresh catalyst (70%). This is a well known effect in zeolite catalysis due to the partial deactivation of the acid sites by carbonaceous deposits. The authors explain these results by a *product selectivity mechanism* inside the pore system of ZSM-5 or ZSM-11. Even though the diffusion pathways for the product molecules decreased by the formation of carbonaceous deposits, the narrow molecules such as 2-MN, 2,6-DMN, and 2,7-DMN were less affected than bulkier ones such as 1-MN and 1,6-DMN.

Yashima and his co-workers examined the methylation of 2-MN using MFI-metallosilicates.[60] The methylation over HZSM-5 yielded predominantly 2,6-, 2,7-, and 2,3-DMN with the formation of bulky 1,2-, 1,3-, 1,6-, and 1,7-DMN (Table 3). The isomers produced initially were three β,β-DMN isomers, *i.e.*, 2,6- (50 %) , 2,7- (38 %), and 2,3- (7%) DMN at the short contact time. However, the selectivities for 2,6- and 2,7-DMN decreased with an increase in α,β-DMN isomers, *i.e.*, 1,3-, 1,6-, and 1,7- DMN as the contact time increased. These results suggest that the increase in the selectivity for α,β-DMN is due to the isomerization of initially produced 2,6- and 2,7-DMN inside the pores. The isomerization occurred at the external acid sites because the co-feed of 2,4-dimethylquinoline (2,4-DMQ) enhanced the selectivity for 2,6- and 2,7-DMN.[10] The isomerization decreased by the substitution of Al with Ga and Fe, and was deactivated completely by the co-feed of 2,4-DMQ for Ga- and Fe-MFI. The substitution also enhanced the formation of 2,6-DMN: the 2,6-/2,7- ratio reached 1.78 for Fe-MFI. These differences of β,β-selectivity and 2,6-/2,7- ratio are not due to the changes of pore dimension but to acid strength by the substitution. Weaker acid sites produce 2,6-DMN preferentially. One reason is that the electron density of HOMO of 2-MN at C-6 is significantly higher than that of C-7.

Table 3 *Methylation of 2-methylnaphthalene over MFI-metallosilicates*[a]

Catalyst	*HZSM-5*		*Ga-MFI*		*Fe-MFI*	
2,4-DMQ	*none*	*co-fed*	*none*	*co-fed*	*none*	*co-fed*
2-MN conv.(%)	25.1	9.4	16.9	9.6	13.9	10.5
1-MN yield(%)	13.9	0.0	6.0	0.0	1.1	0.0
DMN yield(%)	9.9	8.6	9.7	8.6	11.0	9.5
DMN fraction(%)						
1,2-	5.5	5.2	2.1	1.7	2.9	0.9
1,3-	4.8	0.56	4.9	0.8	1.8	0.6
1,6-	17.4	0.1	12.6	0.1	4.8	0.1
1,7-	13.9	10.3	9.5	0.2	2.9	0.2
2,3-	5.9	46.3	7.5	10.7	9.9	10.9
2,6-	26.2	46.0	34.5	46.0	49.2	56.0
2.7-	21.6	37.7	27.4	37.7	28.4	31.4
2,6-/2,7-	1.21	1.22	1.26	1.34	1.73	1.78

[a] Reaction conditions: temperature 450 °C; W/F = 2.48; 2,4-DMQ (2,4-dimethylquinoline) 0.0075 ml/h.

The results described above suggest that the methylation of naphthalene over MFI- metallosilicates occurs inside crystalline pores by a *restricted transition-state mechanism*, not with unordered sites at or near external sites. Weaker acid sites preferentially form 2,6-DMN and favor a decease in coke formation.

4.2 Isopropylation of Naphthalene

4.2.1 Catalysis over Typical Zeolites – In the alkylation of naphthalene, α-alkylation occurs in the initial stage because α-positions are more reactive than β-positions. However, the β/α ratio in the product mixture increases with the increase of reaction temperature and time on stream. Figure 8 shows the three reaction paths for producing diisopropylnaphthalene (DIPN) isomers. The reactions are: (1) alkylation, (2) isomerization, and (3) transalkylation. Isomerization and transalkylation accompany the rearrangement of isopropyl groups. The zeolite type and reaction conditions, *e.g.*, temperature and time on stream, usually determine the type of reaction path.[4]

Table 4 summarizes the distribution of DIPN isomers in the isopropylation.[61,62] High selectivity for 2,6-/2,7-DIPN was observed with HM(25) although its catalytic activity was low. The ratio of the DIPN isomers was almost constant during the reaction. On the other hand, the product distribution over HY and HL was changed by the reaction conditions. The isopropylation with HY selectively formed 1,4-DIPN at low temperatures, particularly at the initial stage of the reaction. However, 1,4-DIPN isomerized to the thermodynamically more stable isomers, *i.e.*, 1,3-, 2,6-, and 2,7-DIPN at high temperatures and/or at prolonged reaction times. The activity of HZSM-5 at 240 °C was extremely low. Song and Kirby reported the similar results in the isopropylation of naphthalene with *i*-PrOH.[63]

Chu and Chen also investigated the isopropylation using *i*-PrOH using flow type reactor at 200 °C under atmospheric pressure.[64] They found that HM and

Table 4 *Isopropylation of naphthalene over typical zeolites*[a]

Catalyst (SiO_2/Al_2O_3)	*Reaction temperature* (°C)	*Reaction period* (h)	*Conv.* (%)	*DIPN yield* (%)	*Distribution of DIPN (%)* 1,3-	1,4-	1,5-	1,6-	1,7-	2,6-	2,7-
HM(25.3)	240	2	35.6	5.7	7.3	5.1	3.3	4.8	3.8	51.7	24.0
	240	8	68.3	18.4	5.3	3.8	1.9	7.1	6.1	50.8	24.9
HY(7.3)	240	2	94.8	43.9	32.3	14.4	8.2	14.7	13.5	8.5	8.4
	240	8	96.1	47.7	23.7	0.6	0.2	6.8	4.9	32.6	31.2
	160	2	90.0	40.5	15.1	39.8	13.1	11.0	15.5	2.8	2.7
	160	8	94.8	45.4	20.0	36.0	12.7	11.6	13.0	3.3	3.4
HL(6.1)	240	2	92.2	42.5	23.1	32.1	10.0	11.2	16.2	3.8	3.4
	240	8	95.1	42.5	39.9	7.9	6.7	15.3	16.3	6.7	7.2
	160	2	66.6	23.5	14.4	38.0	12.4	10.1	18.6	2.9	2.7
	160	8	88.2	39.5	16.6	40.3	11.9	9.6	15.9	3.1	2.6
HZSM-5(70)	240	8	1.0	–	–	–	–	–	–	–	–

[a] Reaction conditions: naphthalene, 0.27 mol; propylene, 0.52 mol; solvent, undecane, 240ml; catalyst, 7 g.

Hβ catalysts had higher selectivities for 2,6-DIPN than HUSY and HY, and the latter catalysts, especially HUSY exhibited the highest activity and stability of all catalysts tested.

Horsley and his co-workers simulated the diffusion of 2,6- and 2,7-DIPN in the HM and HL pores.[65,66] They calculated minimum energy profiles for their diffusion in HM pores. The diffusion of 2,6-DIPN, with a calculated energy barrier of 4 kcal/mol, is significantly less hindered than that of 2,7-DIPN, with a calculated energy barrier of 18 kcal/mol. However, there is no significant barrier for either of the isomers against their diffusion in HL pores. Although factors other than the diffusion may also contribute to the selectivity, these differences in diffusion rates reflect the formation of the transition states. The transition state for the formation of 2,6-DIPN from 2-IPN would be "linear", whereas that of 2,7-DIPN can reasonably be assumed to have a somewhat "bent" configuration, like 2,7-DIPN itself. On the basis of minimum energy profiles for DIPN isomers in HM pores, the transition state for 2,7-DIPN would have a higher energy than that for 2,6-DIPN, resulting in a higher activation energy for the formation of 2,7-DIPN. However, there may be no energy difference in the isopropylation over HL, resulting in non-regioselective formation of 2,6- and 2,7-DIPN.

4.2.2 Effect of the Dealumination of H-Mordenite – The catalytic activities of the isopropylation of naphthalene are not proportional to the acid density and are not enhanced by the dealumination, as is the case of the isopropylation of biphenyl.[18,21,22] Figure 9 shows the yields of products and the distributions of 2-IPN and DIPN isomers at 80% conversion in the alkylation of naphthalene over HM with SiO_2/Al_2O_3 ratio from 10 to 220 at 250 °C under constant propylene pressure of 0.8 MPa.[67,68] The catalytic features of HM with low and high ratios were also quite different. Significant changes in the selectivities for

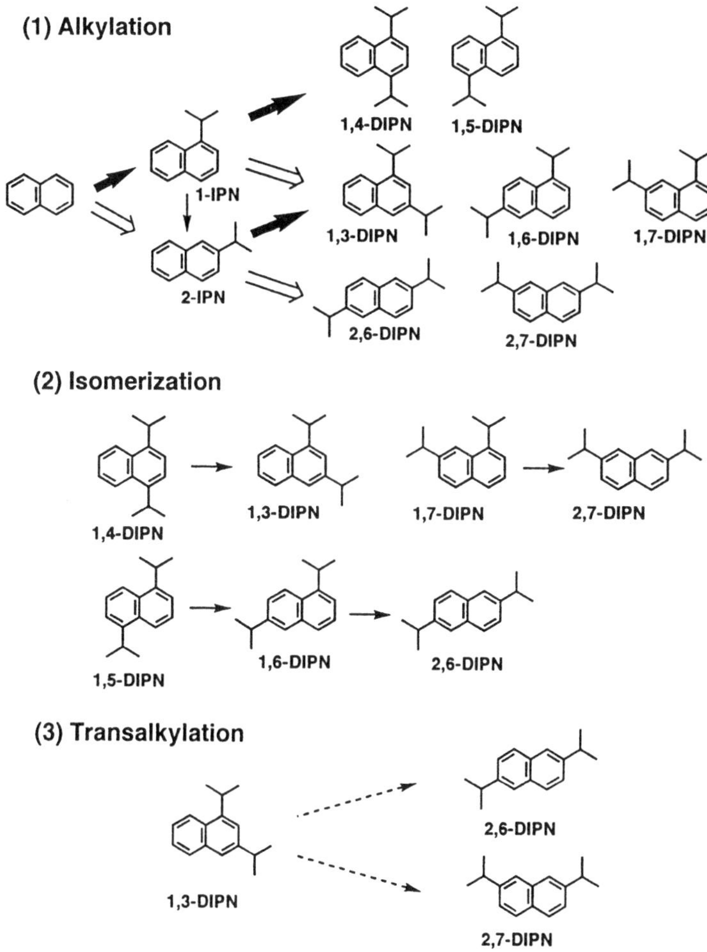

Figure 8 *Reaction path of the isopropylation of naphthalene*

IPN and DIPN isomers occurred at the ratios between 23 and 30. The yield of DIPN isomers and the selectivity for 2,6-DIPN increased rapidly with increasing ratio higher than 30 in spite of a decrease in acid sites by dealumination. The selectivities for β,β-DIPN isomers, especially 2,6-DIPN, were as high as 65% over HM(220). The selectivity for 2,6-DIPN was almost constant during the reaction. However, The catalytic activity and the selectivity of 2,6-DIPN were low for the ratio below 23. The isopropylation over HM(10) was much slower than that over HM(220), and proceeded with a complex change in product distribution. The selectivities for 2,6- and 2,7-DIPN over HM(10) were much lower than those of HM(220). The formation of bulky α,β-isomers was extensive in the early stage. However, these isomers were isomerized to thermodynamically more stable β,β-isomers such as 2,6- and 2,7-DIPN as reaction proceeded. The formation of polyisopropylnaphthalenes (PIPN), such as triisopropylnaphthalenes (TrIPN), and tetraisopropylnaphthalenes (TeIPN), was enhanced over HM(10) as the reaction proceeded. These results indicate that the alkylation over highly dealuminated HM such as HM(220) occurs inside the pores to yield β,β-isomers, especially the 2,6-DIPN, by shape-selective catalysis. On the other hand, the non-regioselective reaction over HM with the low SiO_2/Al_2O_3 ratio occurs mainly at the external surface because most of the acid sites in the pores are not catalytically active sites after they are blocked by coke deposition. A similar phenomenon was also observed in the isopropylation of biphenyl.[21,22]

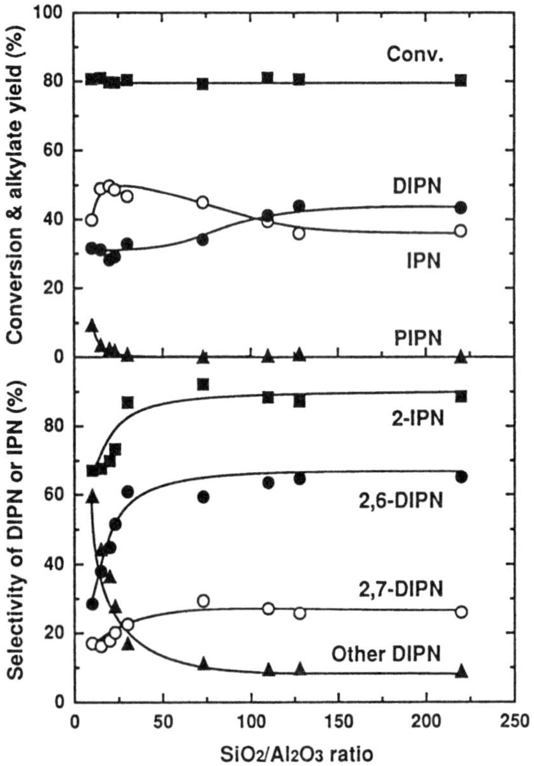

Figure 9 *Effect of the dealumination of HM on the isopropylation of naphthalene. Reaction conditions: naphthalene, 400 mmol; HM, 5 g; propylene pressure, 0.8 MPa; temperature, 250 °C.*

The extent of the coke deposition on HM with a high SiO_2/Al_2O_3 ratio, measured by TG analysis, was apparently different from that on HM with a low ratio.[68] H-mordenite with the ratio higher than 73 had a new type low temperature peak around 350 °C. This peak is ascribed to organic compounds encapsulated inside the pores. These features are quite similar to the case of biphenyl.[25,26] The dealumination causes a decrease in acid density and strength so as to reduce the amount of coke deposition. These property changes result in lowering coke deposition and the enhancement of catalytic activity. A high-silica HM such as HM(220) effectively catalyses shape-selective isopropylation with the minimum coke deposition. Figure 10 shows the effect of the reaction time on coke deposition over HM(10) and HM(220).[69] Over HM(10), the peak around 600 °C due to coke combustion appeared after a short period such as 1 h. The peak becomes progressively greater with the increase of the period. These facts indicate that the pores of HM(10) are choked by coke deposition in short periods. On the other hand, the profiles for HM(220) had a broad peak at around 350 °C and only small peak at 600 °C after the reaction of 1 h. The peak at around 350 °C, which is due to the encapsulated product inside the pores, decreased with time, and the peak due to the coke deposition increased. The different behavior of coke deposition between HM(10) and HM(220) shows that acid density and strength are important factors for the dehydrogenation of naphthalene derivatives, especially PIPN. The composition of PIPN encapsulated in the catalyst was much higher than that of the bulk products although naphthalene was also found in the pores of HM with the low SiO_2/Al_2O_3 ratio. These compounds cannot easily diffuse inside the pores, and are considered to be precursors of deposited coke because the dehydrogenation is easier than that of parent naphthalene or propylene oligomers.

The ratio of 2,6- and 2,7-DIPN encapsulated inside the pores was higher than that of bulk products in the isopropylation of naphthalene.[68] The isopropylation over HM(10) gave principally DIPN isomers other than 2,6- and 2,7-DIPN, and the selectivities for 2,6- and 2,7-DIPN were only 25% and 10%. However, the selectivity of 2,6-DIPN inside the pores of HM(10) after the reaction was as high as 55%. These results support the hypothesis that alkylation inside the pores occurred shape-selectively to yield the narrow product even over HM with a low SiO_2/Al_2O_3 ratio, and that the low selectivity for 2,6- and 2,7-DIPN was due to the alkylation at external surface after choking the pores by coke deposition. Kikuchi and his co-workers also found the enhancement of the selectivity for 2,6-DIPN by the dealumination of HM.[69] They explained that the increase of the selectivity was due to the reduced contribution of the external acid sites because they found great discrepancy between surface bulk SiO_2/Al_2O_3 ratio after the dealumination of their HM. Fellman and his co-workers also proposed the improvement of catalyst performance of HM by the dealumination.[70] These features are quite similar to the case of the isopropylation of biphenyl.[21,22]

We suggest that the selectivity for the isopropylation of naphthalene is explained by *restricted transition-state mechanism*.[68] However, Kikuchi and his co-workers proposed that the selectivity is due to a *product selectivity mech-*

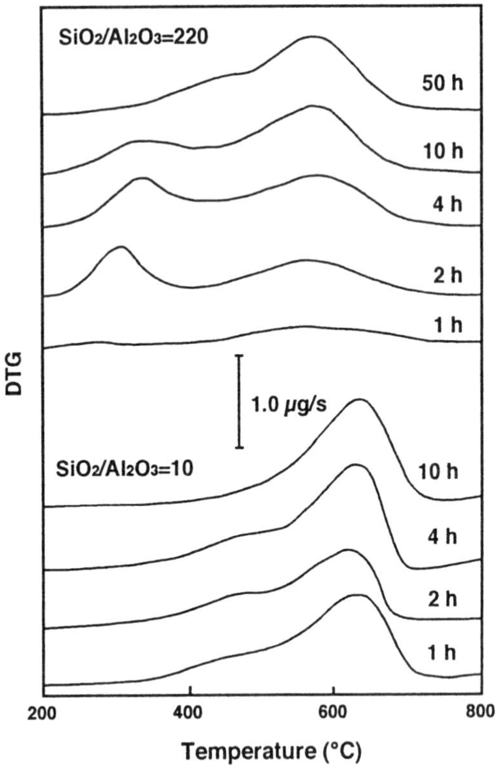

Figure 10 *Effect of reaction period on coke deposition in the isopropylation of naphthalene. Reaction conditions were the same as in Figure 9.*

anism.[69] If the latter mechanism is operated, the compositions of encapsulated DIPN isomers should be close to equilibrium, or at least the selectivities for 2,6- and 2,7-DIPN inside the pores should be low. However, the composition of encapsulated products showed that selective formation of 2,6-DIPN occurred inside the pores. These results suggest that the shape-selective catalysis is explained by steric requirement of the transition state composed of substrate and acid sites in the pores.

Schmitz and Song investigated the relationships between catalyst performance and properties by the dealumination of the same starting H-mordenite, HM(14).[71,72] The catalytic activity was influenced by the extent of dealumination. It increased initially up to a SiO_2/Al_2O_3 ratio of 71, but decreased extensively between 71 and 110. HM(230) had higher activity in spite of the deep dealumination. This type of activity trend, which is a volcano plot for HM catalysis, is due to decreasing number of acid sites and to increasing acid strength by the dealumination; the severe loss of the activity at the high SiO_2/Al_2O_3 ratios is due to the depletion of acid sites. The selectivity for 2,6-DIPN increases from 30% to 60% by dealumination, and the formation of TrIPN fell down to near zero. This indicates that a larger fraction of naphthalene and 2-IPN reacts within the confines of HM pores where α-substitution is sterically blocked. They also observed similar activity changes in the isopropylation of biphenyl.[72] XRD observation showed that the dealumination caused unit cell contraction. They pointed out a slight shrinkage in the channel diameter by the dealumination results in closer fit of 2,6- and 2,7-DIPN in the channels, and that diffusion of 2,7-DIPN becomes even more hindered than in the non-dealuminated HM case.

Schmitz and Song investigated the reactivity of 2,6-DIPN over HM under propylene pressure.[73] Over HM(14) and HM(38), 2,6-DIPN conversions were 12-27% mostly to higher alkylates. The isopropylation of 2,6-DIPN increased over these HM under high propylene pressures, but its isomerization was suppressed. Both reactions of 2,6-DIPN produce α-substituted naphthalenes by non-regioselective reactions on the external surface. On the other hand, 2,6-DIPN was nearly unreactive over HM(74) because of fewer external acid sites compared with HM(14) and HM(38). These results reveal that high selectivity toward 2,6-DIPN in the isopropylation of naphthalene is partly ascribed to the low reactivity of 2,6-DIPN.

4.2.3 Deactivation of External Acid Site of H-Mordenite – It is important to deactivate the external acid sites because the reactions on them are more rapid than that of inside pores, causing lower selectivity for 2,6-DIPN. The acidity measured by the cracking of 1,3,5-TIPB showed that HM is active even after the dealumination in spite of the decrease of the number of acid sites on external surface.[68] These results did not always correspond to catalytic activity of the isopropylation of naphthalene, but a small number of acid sites on the external surface over HM(128) is sufficient for the cracking reaction. The present authors examined the deactivation by the impregnation of ceria on the HM.[50,51] Figure 11 shows the effect of cerium modification on the selectivity for 2,6-DIPN in the isopropylation of naphthalene at 250 °C under propylene pressure of 0.8 MPa. The selectivity for 2,6-DIPN decreased with operating time in the isopropylation over HM(128). However, the selectivity for 2,6- and 2,7-DIPN over Ce(30)HM(128)* was almost constant at 70% during the reaction. This value is one of the highest selectivities for 2,6-DIPN reported. An amount of cerium higher than 30 wt% of the HM was required for the improvement of the selectivity for 2,6-DIPN, and low loading of cerium did not necessarily improve the selectivity significantly. High loading of cerium, 30-50%, should be necessary for covering external acid sites with keeping open HM pores because of heavy atomic weight of cerium in comparison with silicon and aluminum. These results show that the modification with ceria selectively deactivated acid sites for the isomerization of 2,6-DIPN to 2,7-DIPN or for non-regioselective alkylation without blocking the HM pores.

* The number in the parentheses after "Ce" expresses cerium amount (wt % to the HM).

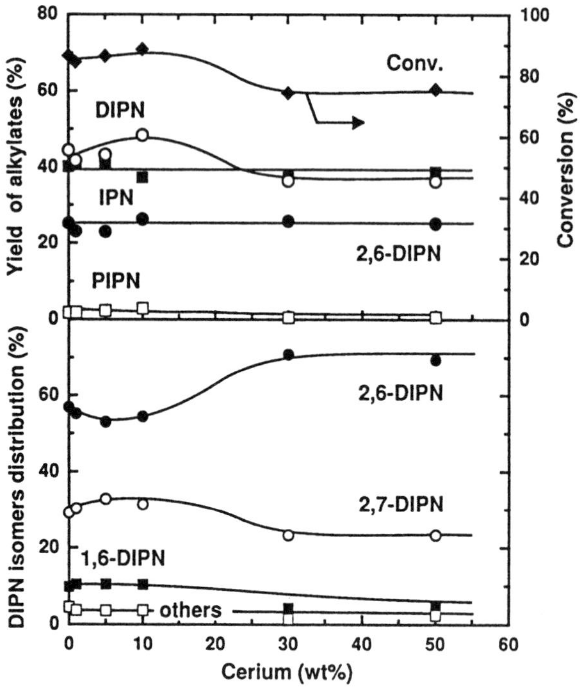

Figure 11 *The effect of cerium amount on the isopropylation of naphthalene over ceria modified H-mordenite. Reaction conditions: catalyst, 1 g; naphthalene, 200 mmol; propylene pressure, 0.8 MPa; temperature, 300 °C; period, 4 h.*

It is important to calcine at the appropriate temperature to enhance the deactivation and the catalysis for the ceria-modified HM.[50,51] The catalyst calcined at 300 °C had no catalytic activity. The activity for the isopropylation of naphthalene appeared over the catalyst calcined at 450 °C and reached the maximum at 550 °C. However, the activity decreased at a temperature of 700 °C. These results show that calcination at an appropriate temperature is necessary to keep the HM pores open. The selectivity for 2,6-DIPN was as high as 70% over the catalysts regardless of their calcination temperature. These phenomena show that the isopropylation proceeds inside HM pores. This selective deactivation of HM pores was limited to modification with ceria. The modification with other rare-earth metal oxides, such as lanthanum and neodymium oxides, also deactivated the external acid sites but choked the HM pore to result in low catalytic activity. The isomerization of 4,4′-DIPB was also prevented by the ceria modification of HM in the isopropylation of biphenyl.[49]

Moreau and his co-workers examined CVD silanation of HM and HY with octamethylcyclotetrasiloxane for the deactivation of external acid sites.[74] They found the decrease of the formation of TrIPN in the isopropylation of naphthalene with isopropyl bromide was because of the deactivation of external acid sites although the selectivity for 2,6-/2,7-DIPN was not changed.

Schmitz and Song found that the selectivity for β,β-DIPN, especially 2,6-DIPN, increased by adding water to the catalysts, HM(38) and HM(74).[75] Without the addition of water for HM(38) and HM(74), β,β-selectivities were 56% and 77%, respectively, and the selectivities for 2,6-DIPN were 37% and 54%, respectively. With the addition of water, β,β-selectivities increased to 90% and 95%, and 2,6-DIPN selectivities for HM(38) and HM(74) increase to 60% and 70%, respectively. A large amount of water always decreased the activity, but mixed effects were observed for intermediate amounts of added water. Coke deposits also decreased with the addition of water. Sorption of water on the catalysts could explain changes of the selectivity and the activity. They suggest that a small amount of water is sufficient to deactivate non-regioselective reaction at external acid sites and that the increase of the selectivity is due to the formation of new type of Brønsted acid sites located inside the HM pore by dissociative adsorption of water. The higher β,β-selectivities obtained with *i*-PrOH as an alkylating agent, rather than with propylene, could be due to water formed from the alcohol.

4.2.4 Isopropylation over HY, HL and Other Zeolites – The catalysis over HY and HL was quite different from that over HM.[61-63] The further isopropylation of IPN and DIPN isomers was quite rapid in spite of the higher number of bulky isopropyl groups. The yield of IPN had a maximum value during the initial stage of the reaction, and then decreased with reaction time. At longer periods, the product amounts decreased as follows: DIPN > IPN > TrIPN. Because the catalytic activities correlate well with the pore surface areas of the zeolite, the isopropylation occurs predominantly inside the pores. These results indicate that HY and HL have enough space for the isopropylation of naphthalene inside their pores. Naphthalene and its alkylates can easily enter into and diffuse out of the pores.

HY and HL gave predominantly α,α-DIPN such as 1,4-DIPN at the initial reaction stage.[4,61,62] However, α,α-DIPN changed to the more thermodynamically stable α,β- and β,β-DIPN (1,3-, 2,6-, and 2,7-DIPN) at prolonged periods. As shown in Figure 12, the composition of DIPN isomers versus reaction period is dependent on zeolite type and reaction conditions.[4] At 240 °C over HY, 1,4-DIPN was the initial product, and the selectivity was up to 40%, but it decreased by the isomerization to 1,3-DIPN. These compounds converted to approximately equal amounts of 2,6- and 2,7-DIPN after prolonged periods. However, 1,4-DIPN was obtained in high yield at the low temperature of 160 °C. These changes reflect the differences in reactivity and thermodynamic stability of the α-positions of naphthalene. The product distribution for HY was governed kinetically at low temperatures and/or at a short reaction period, but thermodynamically at high temperatures. The isopropylation occurs at active α-positions to form α,α-DIPN in the first step. In the second step, α,α-DIPN converted to thermodynamically stable α,β- and β,β-DIPN by transalkylation and dealkylation-realkylation at acid sites in the supercages of HY.

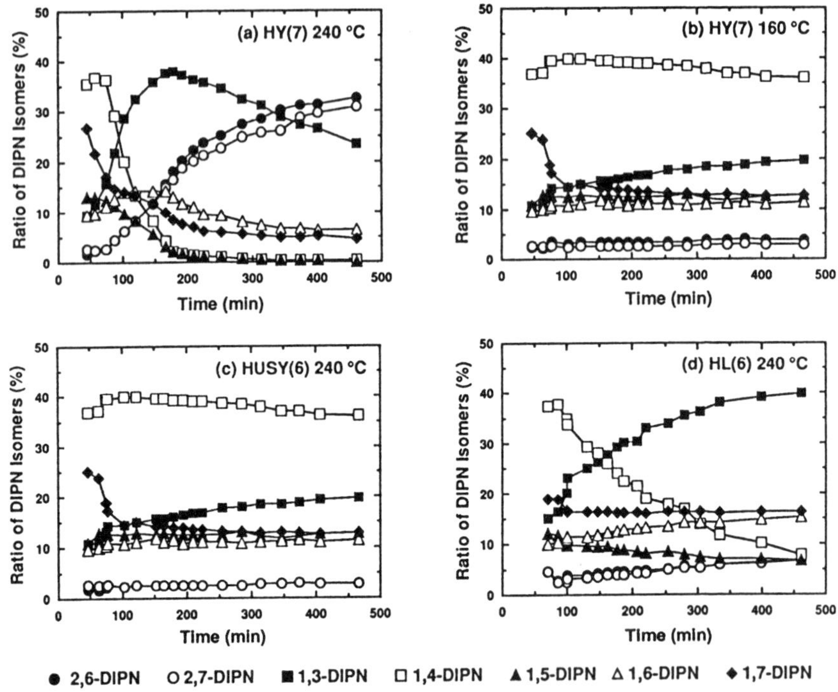

Figure 12 *Reaction period vs. the distribution of DIPN isomers in the isopropylation of naphthalene over HY and HL. Reaction conditions: naphthalene, 0.27 mmol; propylene, 0.48 mol; catalyst, 7 g; solvent, undecane, 240 ml; temperature 240 °C.*

Chu and Chen described the shape-selective synthesis of 2,6-DIPN in the isopropylation of naphthalene over Hβ with *i*-PrOH using a flow-type reactor.[64,76] The selectivity for 2,6-DIPN increased with decreasing the SiO_2/Al_2O_3 ratio, although the activity decreased. Up to 80% selectivity for 2,6-DIPN was obtained for ratios above 110. These results are explained by lowering acid density and increasing particle size of corresponding Hβ with increasing SiO_2/Al_2O_3 ratio.

4.2.5 Alkylation with Other Alkenes – Fellman pointed out that the composition of 2,6-diethylnaphthalene (2,6-DEN) and the 2,6-/2,7-DEN ratio is an important indicator for the selective ethylation of naphthalene.[77] The selective ethylation requires the 2,6-DEN composition to exceed thermodynamic equilibrium and the 2,6-/2,7-DEN ratio to be higher than 1.0. Takeuchi and his co-workers confirmed experimentally over SA, HY, and HM that all of 2,6-DEN compositions were less than 27%, which was the estimated value at thermodynamic equilibrium, although the ratios were higher than 1.0.[78] Other DEN isomers having at least one ethyl group at the α-position were formed in higher than 20% in the ethylation over all catalysts. These results suggest that the ethylation of naphthalene is not shape-selective even by the use of zeolites, which have a smaller pore size than the mordenite. The discrepancy between the ethylation and the isopropylation could be due to the difference of the corresponding transition states.

Moreau and his co-workers showed that the higher alkylating reagent, cyclohexene or cyclohexyl bromide, gave β,β-dicyclohexylnaphthalenes selectively over HY.[79-81] However, the 2,6-/2,7- ratio was almost unity.

5 Alkylation of Other Polynuclear Hydrocarbons

The isopropylation of *p*-terphenyl (TP) was similar to the isopropylation of biphenyl.[82] Over HM, the isomer compositions of 4-isopropyl-*p*-terphenyl (4-IPTP) and 4,4″-diisopropyl-*p*-terphenyl (4,4″-DIPT) at 175~200 °C are 78 and 75%, respectively. The selectivity for 4,4″-DITP was higher than that for 4,4′-DIPB under comparable reaction conditions. This difference is possibly caused by the differences in the steric restriction and the diffusion rates between both reactions. However, the isopropylation over SA was non-regioselective: The selectivities were 35~50% and 10~20% for 4-IPTP and 4,4″-DIPT, respectively.

The isopropylation of dibenzofuran (DBF) over HM at temperatures below 200 °C produced 2-isopropyldibenzofuran (2-IPDBF) in preference to other IPDBF isomers.[83] Since the molecular sizes of 1- and 4-IPDBF are larger than those of 2- and 3-IPDBF, the formation of 1- and 4-IPDBF is suppressed because the corresponding bulky transition state requires more space than is available at the acid sites. However, the most reactive positions of dibenzofuran are the 2- and 4- positions because of their high electron densities. The overall selectivity for 2-IPDBF over HM is due to the synergism between the environments of transition states in the pore and the differences in the reactive positions of dibenzofuran. On the other hand, HY and SA did not exhibit shape-selectivity in isopropylation. The selectivities for 1-, 2-, 3-, and 4-IPDBF over HY at 150 °C were 23, 32, 17, and 28%, respectively. The initial amounts of these four IPDBF isomers over SA at 150 °C decreased in the order of 2-IPDBF >> 4-IPDBF > 1-IPDBF >> 3-IPDBF, although 2- and 1-IPDBF decreased at thermodynamic equilibrium. However, 3- and 4-IPDBF increased, with the overall amounts decreasing in the order of 4-IPDBF > 3-IPDBF > 2-IPDBF > 1-IPDBF.

The isopropylation of anthracene gave similar results to that of naphthalene.[84] The selectivities for 2-isopropylanthracene (2-IPA) and 2,6-diisopropylanthracene (2,6-DIPA) over HM(25) were as high as 91% and 47%, respectively. On the other hand, the selectivities over HY were as low as 59% and 8% for both products. The Diels-Alder reaction of anthracene with propylene occurred at higher temperatures to yield large amounts of the adducts such as 11-methyl-9,10-dihydro-*etano*-anthracene. However, their formation was prevented by the addition of a small quantity of water.

6 Transalkylation of Biphenyl and Naphthalene

The ethylation of biphenyl with ethylene even over HM did not give the high selectivity for 4,4′-DEBP as discussed previously.[54,55] The lack of the selectivity lies in the size of the ethyl group because the HM pore does not exert its shape-selective catalysis for a less bulky group than the isopropyl group. However, the selective formation of 4,4′-DEBP in high selectivity (86%) and good yield (53%) was achieved by the transethylation of biphenyl with tri- or tetraethyl benenes over HY at 170 °C.[56] These high selectivities for 4,4′-DEBP are not ascribed to the shape-selectivity of HY, but to the effective modification of transethylation conditions.

The ethylation of naphthalene was not shape-selective even over HM. However, the transethylation of naphthalene with polyethylbenzenes over HY gave high selectivity for 2,6-DEN with high yield.[78] Tetraethylbenzene was the best transalkylating agent among di-, tri-, and tetraethylbenzenes to form 2,6-DEN. The highest selectivity and yield of 2,6-DEN were 62.4 and 46.6%, respectively, at 110 °C for 8 h. Because the silica-alumina catalyst yielded 2,6-DEN in high selectivity, the transethylation was not explained by the shape-selective catalysis, either. The key factor for the high selectivity is the steric hindrance and high electron density at the substitution site of naphthalene.

7 Conclusion

The product distribution in the zeolite-catalysed alkylation of polynuclear aromatics depends on the structure of zeolite pores. High regioselectivities were observed in the HM catalysed isopropylation of polynuclear aromatics, such as biphenyl, naphthalene, *p*-terphenyl, and dibenzofuran, to yield predominantly the least bulky products: *e.g.*, 4,4′-DIPB for biphenyl, and 2,6-DIPN for naphthalene. These reactions are controlled by steric restriction at the transition state inside the pores and by the entrance of intermediate products molecules into the pores. On the other hand, the catalysts with large-pore HY and HL zeolite are controlled at low temperature by the electron density of the reactant molecule and at higher temperature by the stability of the product molecules because their pores have enough space for a transition state, which allow the formation of all corresponding isomers.

Dealumination of HM resulted in common features for both the isopropylation of biphenyl and naphthalene. The dealumination of HM decreases the blocking of the pores by coke deposition in the both reactions because of the decrease in acid sites at intracrystalline and external surfaces. It also reduced the non-regioselective isopropylation at the external surface, and enhanced the reaction to yield the least bulky product molecules. The pressure of propylene is also an important key factor for high yield of 4,4′-DIPB in the isopropylation of biphenyl, because the isomerization of 4,4′-DIPB occurred significantly under low pressure of propylene at the external acid sites.

H-mordenite has active external acid sites for acid catalysis even after the deep dealumination. It is important to deactivate the external acid sites because the reaction on them is more rapid and less regioselective than that inside the pores. Modification with ceria, silica, and phosphorus oxide could selectively deactivate the external acid sites in the isopropylation of naphthalene and biphenyl. In particular, modification with ceria and phosphorus oxide improved effectively the shape-selective catalysis.

The pores of HZSM-5 do not have enough space to accommodate 1-MN and DMN isomers with α-methyl groups. The methylation of 2-MN over HZSM-5 yields predominantly β,β-DMN isomers, 2,6-, 2,7- and 2,3-DMN, with the formation of bulky α,β-DMN, 1,2-, 1,3-, and 1,6-, and 1,7-DMN. The selective formation of 2,6- and 2,7-DMN occurs inside pores, because the deactivation of external acid sites of MFI-metallosilicates by 2,4-DMQ reduces the isomerization of 2,6- and 2,7-DMN to α,β-DMN. However, the pores of HM and HY do not contribute to the shape-selective methylation and related reactions of naphthalene and 2-MN.

The zeolite-catalysed synthesis of symmetrically dialkylated polynuclear aromatics is a promising way to synthesize advanced materials. To achieve high selectivity and catalytic activity, it is essential to design the zeolite pores and the stereochemistry of the transition state, which is composed of the aromatic reactant, the alkylating agent, and the acid sites inside the pore. The investigation should be focused on the following points: (1) the minimization of steric restriction of transition states composed of polynuclear aromatics, alkylating agent and acids sites in the zeolite pore; (2) the prevention of the reactions at the external surface; (3) the control of number of acid sites inside the pores; and (4) the control of acid strength.

Acknowledgments

The authors thank to Tosoh Corporation for the gift of zeolites used in this work, to Professor Chun-shan Song of Pennsylvania State University, and Professor Seitaro Namba of Teikyo University of Science & Technology for their valuable discussions, and to Drs. T. Matsuzaki, T. Hanaoka, and X. Tu of National Institute of Materials and Chemical Research, and Professor J.-H. Kim of Chonnam National University for their continuing cooperation.

A part of this work was supported by New Energy and Industrial Technology Development Organization (NEDO) of Japan.

References

1. S.M. Csicsery, *Zeolite,* 1984, **4**, 202.
2. J. Weitkamp and S. Ernst, *Catal. Today,* 1994, **19**, 107.
3. C.B. Dartt and M.E. Davis, *Catal. Today,* 1994, **19**, 151.
4. Y. Sugi and M. Toba, *Catal. Today,* 1994, **19**, 187.
5. P.B. Venuto, *Micropor. Mater.,* 1994, **2**, 297.
6. C. B. Khouw and M.E. Davis, in "Selectivity in Catalysis" (ACS Symposium Series,

517), ed. M.E. Davis and S.L. Suib, American Chemical Society, Washington DC, 1993, p. 207.
7. J. Fellmann, *Catalytica Highlights,* 1991, **17**(#1), 1.
8. A. Cortes and A. Corma, *J. Catal.,* 1971, **51**, 338.
9. W.W. Kaeding, C. Chu, L.B. Young, B. Weinstein, and J. Butter, *J. Catal.,* 1981, **67**, 159.
10. S. Namba, J.-H. Kim, and T. Yashima, in "Zeolites and Microporous Crystals" (Stud. Surf. Sci. Catal., vol. 83), ed. T. Hattori and T. Yashima, Kodansha-Elsevier, 1994, p. 279 and their earlier papers cited in.
11. L.B. Young, S.A. Butter, and W.W. Kaeding, *J. Catal.,* 1982, **76**, 418.
12. H.G. Karge, J. Laderbeck, Z. Sarbak, and K. Hatada, *Zeolites,* 1982, **2**, 94, and their earlier papers.
13. S.M. Csicsery, *J. Catal.,* 1971, **23**, 124.
14. E.G. Derouane, *J. Catal.,* 1986, **100**, 541.
15. E.G. Derouane, J.-M. Andre, and A.A. Lucas, *J. Catal.,* 1988, **110**, 58.
16. E. G. Derouane and Z. Gabelica, *J. Catal.,* 1980, **65**, 486.
17. T. Matsuzaki, Y. Sugi, T. Hanaoka, K. Takeuchi, H. Arakawa, T. Tokoro, and G. Takeuchi, *Chem. Express,* 1989, **4**, 413.
18. Y. Sugi, T. Matsuzaki, T. Hanaoka, K. Takeuchi, T. Tokoro, and G. Takeuchi, in "Chemistry of Microporous Crystals" (Stud. Surf. Sci, Catal., vol. 60), ed. T. Inui, S. Namba, and T. Tatsumi, Kodansha-Elsevier, Tokyo-Amsterdam, 1991, p.303.
19. Y. Sugi, T. Matsuzaki, T. Tokoro, T. Hanaoka, K. Takeuchi, X. Tu, and G. Takeuchi, *Sekiyu Gakkaishi,* 1994, **37**, 376.
20. Y. Sugi, T. Matsuzaki, T. Hanaoka, Y. Kubota, J.-H. Kim, X. Tu, and M. Matsumoto, *Catal. Lett.,* 1994, **26**, 181.
21. Y. Sugi, T. Matsuzaki, T. Hanaoka, Y. Kubota, J.-H. Kim, X. Tu, and M. Matsumoto, in "Acid-Base Catalysis II" (Stud. Surf. Sci. Catal., vol. 90), ed. H. Hattori, M. Misono, and Y. Ono, Kodansha-Elsevier, Tokyo-Amsterdam, 1994, p. 391.
22. X. Tu, M. Matsumoto, T. Matsuzaki, T. Hanaoka, Y. Kubota, and Y. Sugi, *Catal. Lett.,* 1993, **21**, 71.
23. X. Tu, M. Matsumoto, T. Maeda, Y. Sugi, T. Matsuzaki, Y. Kubota, and J.-H. Kim, in "Advanced Materials '93, II/A: Biomaterials, Organic and Intelligent Materials", ed. H. Aoki, K. Segawa, T. Nishi, M. Hasegawa, I. Karube, T. Kitajima, and K. Takahashi, Elsevier, Amsterdam, 1994, p. 157.
24. Y. Sugi, X. Tu, T. Matsuzaki, T. Hanaoka, Y. Kubota, J.-H. Kim, M. Matsumoto, K. Nakajima, and A. Igarashi, *Catal. Today,* 1996, **31**, 3.
25. Y. Sugi, T. Matsuzaki, T. Hanaoka, Y. Kubota, J.-H. Kim, X. Tu, and M. Matsumoto, *Catal. Lett.,* 1994, **27**, 315.
26. Y. Sugi, T. Matsuzaki, T. Hanaoka, Y. Kubota, J.-H. Kim, X. Tu, and M. Matsumoto, *Sekiyu Gakkaishi,* 1994, **37**, 557.
27. Y. Sugi, X. Tu, J.-H. Kim, T. Matsuzaki, T. Hanaoka, Y. Kubota, and M. Matsumoto, *Prepr. Div. Petrol. Chem. Am. Chem. Soc.,* 1995, **40**, 576.
28. Y. Sugi, X. Tu, T. Matsuzaki, T. Hanaoka, Y. Kubota, J.-H. Kim, M. Matsumoto, K. Nakajima, and A. Igarashi, *Shokubai,* 1995, **37**, 152.
29. T. Hanaoka, K. Nakajima, T. Matsuzaki, Y. Kubota, Y. Sugi, A. Igarashi, and K. Kunimori, *Catal. Lett.*, in press.
30. W.M. Meier, D.H. Olson, and Ch. Baerlocher, in "Atlas of Zeolite Structure Types", 4th revised edition, Elsevier, Amsterdam, 1996.
31. P. Dejaifve, A. Auroux, P. C. Gravelle, J. C. Vedrine, Z. Gabelica, and E. Derouane, *J. Catal.,* 1981, **70**, 123.
32. M. Neuber, S. Ernst, P.J. Gobet, P.A. Jacobs, G.T. Kokotailo, and J. Weikamp, in "Catalyst Deactivation 1987" (Stud. Surf. Sci. Catal., vol. 34) ed. B. Delmon and G.F. Froment, Elsevier, Amsterdam, 1987, p. 567.
33. M. Neuber, H.G. Karge, and J. Weitkamp, *Catal. Today,* 1988, **3**, 11.
34. M. Guisnet and P. Magnoux, *Appl. Catal.,* 1989, **54**, 1.

35. S. Bhatia, J. Beltramini, and D.D. Do, *Catal. Rev. Sci,. Eng.*, 1989-90, **31**, 431.
36. H. G. Karge and J. Weitkamp, *Chem. Ind. Tech.*, 1986, **58**, 946.
37. L. Meyers, T.H. Fleisch, G.J. Ray, J.T. Miller, and J.B. Hall, *J. Catal.*, 1988, **110**, 82.
38. M. Sawa, M. Niwa, and Y. Murakami, *Appl. Catal.*, 1989, **53**, 169, and their earlier papers cited therein.
39. G.S. Lee, J.J. Maj, S.C. Rocke, and J.M. Garces, *Catal. Lett.*, 1989, **2**, 243.
40. G.S. Lee, J.M. Garces, and J.J. Maj, in "Catalytic Science and Technology, vol. 1", ed. S. Yoshida, N. Takezawa, and T. Ono, Kodansha-VCH, Tokyo-Weinheim, 1991, p.385.
41. G.J. Lee, J.M. Garces, and J.J. Maj, US Patent, 5015,797 (assigned to Dow Chem. Comp. Ltd.).
42. K. Itabashi, T. Fukushima, and K. Igata, *Zeolites,* 1986, **6**, 30.
43. G. Takeuchi, H. Okazaki, T. Kito, Y. Sugi, and T. Matsuzaki, *Sekiyu Gakkaishi,* 1991, **34**, 242.
44. T. Matsuda, T. Urata, U. Saito, and E. Kikuchi, *Appl. Catal.*, 1995, **131**, 215.
45. T. Taniguchi, M. Tanaka, K. Takahata N. Sakamoto, T. Takai, Y. Kurano, and M. Ishibushi, PCT Int., Appl., WO 88 03523 (assigned to Mitsui Petrochem. Ind. Ltd.).
46. N. Sakamoto, T. Takai, K. Taniguchi, K. Takahata, Jpn. Tokkyo Kokai Koho, 88-122635 (assigned to Mitsui Petrochem. Ind. Ltd).
47. T. Matsuda and E. Kikuchi, in "Zeolites and Microporous Crystals" (Stud. Surf. Sci. Catal. vol. 83), ed. T. Hattori and T. Yashima, Kodansha-Elsevier, Tokyo-Amsterdam, 1994, p. 295.
48. T. Matsuda, T. Urata, E. Kikuchi, *Appl. Catal.*, 1995, **123**, 205.
49. K. Nakajima, T. Hanaoka, T. Matsuzaki, Y. Kubota, Y. Sugi, K. Takeuchi, A. Igarashi, and K. Kunimori, paper in preparation.
50. Y. Sugi, J.-H. Kim, T. Matsuzaki, T. Hanaoka, Y. Kubota, X. Tu, and M. Matsumoto, in "Zeolites and Related Microporous Materials: State of the Art 1994" (Stud. Surf. Sci. Catal., vol. 84) ed. J. Weitkamp, H.G. Karge, H. Pfeifer, and W. Hölderlich, Elsevier, Amsterdam, 1994, p. 1837.
51. J.-H. Kim, Y. Sugi, T. Matsuzaki, T. Hanaoka, Y. Kubota, X. Tu, M. Matsumoto, A. Kato, G. Seo, and C. Pak, *Appl. Catal., A: General,* 1995, **131**, 15.
52. T. Matsuda, T. Kimura, E. Herawati, C. Kobayshi, and E. Kikuchi, *Appl. Catal., A: General,* 1996, **136**, 19.
53. A.S. Loktev and P.S. Chekriy, in "Zeolites and Related Microporous Materials: State of the Art 1994" (Stud. Surf. Sci. Catal., vol. 84) ed. J. Weitkamp, H.G. Karge, H. Pfeifer, and W. Hölderlich, Elsevier, Amsterdam, 1994, p.1845.
54. X. Tu, M. Matsumoto, T. Maeda, Y. Sugi, T. Matsuzaki, T. Hanaoka, Y. Kubota, and J.-H. Kim, *Micropor. Mater.* 1995, **3**, 593.
55. M. Matsumoto, X. Tu, Y. Sugi, T. Matsuzaki, T. Hanaoka, Y. Kubota, J.-H. Kim, K. Nakajima, A. Igarashi, and K. Kunimori, in "Progress in Zeolite and Microporous Materials" (Stud. Surf. Sci. Catal., vol.105), ed. H. Chon, S.-K. Ihm, and Y.S.Uh, Elsevier, Amsterdam, 1997, p.1317.
56. G. Takeuchi, Y. Shimoura, and T. Hara, *Appl. Catal., A: General,* 1996, **137**, 87.
57. D. Fraenkel, M. Cherniavsky, and M. Levy, Proc. 8th Intern. Congr. Catal., Verlag Chemie, Weinheim, 1984, vol. 4, p. 545.
58. D. Fraenkel, M. Cherniavsky, B. Ittah, and M. Levy, *J. Catal.*, 1986, **101**, 273.
59. J. Weitkamp and N. Neuber, in "Chemistry of Microporous Crystals" (Stud. Surf. Sci. Catal., vol. 60), ed., T. Inui, S. Namba, and T. Tatsumi, Kodansha-Elsevier, Tokyo-Amsterdam, 1991, p.291.
60. T. Komatsu, Y. Araki, S. Namba, T. Yashima, in "Zeolites and Related Microporous Materials: State of the Art 1994" (Stud. Surf. Sci. Catal., vol. 84) ed. J. Weitkamp, H.G. Karge, H. Pfeifer, and W. Hölderlich, Elsevier, Amsterdam, 1994, p. 1821.
61. M. Toba, A. Katayama, F. Mizukami, S. Niwa, Y. Sao, and K. Maeda, *Shokubai,* 1990, **32**, 392.

62. A. Katayama, M. Toba, G. Takeuchi, F. Mizukami, S. Niwa, and S. Mitamura, *J. Chem. Soc., Chem. Commun.*, **1991**, 39.
63. C. Song and S. Kirby, *Micropor. Mater.*, 1994, **2**, 467.
64. S.-J. Chu and Y.-W. Chen, *Appl. Catal.*, 1995, **123**, 51.
65. J.A. Horsley, J. D. Fellmann, E. G. Derouane, and C. M. Freeman, in "Computer Aided Innovation of New Materials II", ed. M. Doyama, J. Kimura, M. Tanaka, and R. Yamamoto, Elsevier, Amsterdam, 1993, p. 985.
66. J.A. Horsley, J. D. Fellmann, E. G. Derouane, and C. M. Freeman, *J. Catal.*, 1994, **147**, 231.
67. J.-H. Kim, T. Matsuzaki, K. Takeuchi, T. Hanaoka, Y. Kubota, Y. Sugi, X. Tu, and M. Matsumoto, in "Advanced Materials '93, II/A: Biomaterials, Organic and Intelligent Materials", ed. H. Aoki, K. Segawa, T. Nishi, M. Hasegawa, I. Karube, T. Kitajima, and K. Takahashi, 1994, p.153.
68. J.-H. Kim, T. Matsuzaki, T. Hanaoka, Y. Kubota, Y. Sugi, M. Matsumoto, and X. Tu, *Micropor. Mater.*, 1995, **5**, 113.
69. E. Kikuchi, K. Sawada, M. Maeda, and T. Matsuda, in "Acid-Base Catalysis II" (Stud. Surf. Sci. Catal., vol. 90), ed. H. Hattori, M. Misono, and Y. Ono, Kodansha-Elsevier, Tokyo-Amsterdam, 1994, p. 391.
70. D. Fellmann, R. J. Saxon, P. R. Wentrcek, E. G. Derouane, and P. Massiani, PCT Int., Appl., WO 90 03961 (assigned to Catalytica, Inc.).
71. A.D. Schmitz and C. Song, *Am. Chem. Soc. Div. Fuel Chem. Prepr.*, 1994, **39**, 986.
72. A.D. Schmitz and C. Song, *Am. Chem. Soc. Div. Fuel Chem. Prepr.*, 1995, **40**, 918.
73. A.D. Schmitz and C. Song, *Catal. Today*, 1996, **31**, 19.
74. P. Moreau, A. Finiels, P. Geneste, F. Moreau, and J. Solofo, *J. Catal.*, 1992, **136**, 487.
75. A.D. Schmitz and C. Song, *Catal. Lett.*, 1996, **40**, 59.
76. S.-J. Chu and Y.-W. Chen, *Ind. Eng. Chem. Res.*, 1994, **33**, 3112.
77. J.D. Fellman, P. Massoani, H. Exner, P.R. Wentreck, and R.J. Saxton, PCT Int., Appl., WO 91 05751 (1991) (assigned to Catalytica, Inc.).
78. G. Takeuchi, Y. Shimoura and T. Hara, *Catal. Lett.*, 1996, **41**, 195.
79. P. Moreau, A. Finiels, P. Geneste, F. Moreau, and J. Solofo, *J. Org. Chem.*, 1992, **57**, 5040.
80. P. Moreau, A. Finiels, P. Geneste, F. Moreau, and J. Solofo, *Catal. Today*, 1996, **31**, 11.
81. D. Mravec, M. Michvocik, M. Hronec, P. Moreau, A. Finiels, and G. Genste, *Catal. Lett.*, 1996, **38**, 267.
82. G. Takeuchi, H. Okazaki, M. Yamaye, and T. Kito, *Sekiyu Gakkaishi*, 1991, **34**, 531.
83. G. Takeuchi, H. Okazaki, M. Yamaye, and T. Kito, *Appl. Catal.*, 1991, **76**, 49.
84. F. Mizukami, A. Katayama, S. Mitamura, and M. Toba, Jpn. Tokkyo Kokai, 91-176429 (assigned to Agency of Industrial Science and Technology and Nippon Steel Corp.).

4

Preparation and Characterization of Hexaaluminate Materials for High-temperature Catalytic Combustion

BY G. GROPPI, C. CRISTIANI AND P. FORZATTI

1 Introduction

Catalytic combustion for gas turbines has received much attention in recent years in view of its unique capability of simultaneous control of NO_x, CO, and unburned hydrocarbon emissions.[1] One of the major challenges to be faced in the development of industrial devices is associated with the severe requirements on catalytic materials posed by extreme operating conditions of gas turbine combustors. The catalytic combustor has to ignite the mixture of fuel (typically natural gas) and air at low temperature, preferably at the compressor outlet temperature (about 350 °C), guarantee complete combustion in few milliseconds, and withstand strong thermal stresses arising from long-term operation at temperatures above 1000 °C and rapid temperature transients.

Among the catalysts investigated, hexaaluminate materials (e.g. $BaAl_{12}O_{19}$) are promising because of the unique thermal stability of their phase composition, morphology, and catalytic activity.[2] This property is related to the peculiar layered alumina structure that forms at thermodynamic equilibrium in M-Al-O systems (M is a monovalent, divalent, or trivalent ion of an alkaline, alkaline-earth, or rare earth metal) with Al/M ratio around 12.[3] These structures consist of alternative stacks of Al spinel blocks and mirror planes in which the large cations of the alkaline, alkaline earth, or rare earth metals are located. Two structures are possible (Figure 1): β-Al_2O_3 and magnetoplumbite, which differ in the composition of the mirror planes.[4]

Because of the close similarity of ionic radii, transition metal ions (M = Mn, Cu, Fe, Cr, Co, Ni) can be partially substituted for Al ions. These transition metal ions can provide significant activity in combustion reaction.[5] At low metal concentrations, the structural and morphological properties are not significantly affected by substitution, so that high thermal stability is maintained. Substituted hexaaluminates can be directly shaped in the monolith form required by the combustor, providing bulk active catalysts without need of ceramic supports.

In this chapter, the different preparation methods of hexaaluminates proposed in the literature are reviewed along with the thermal evolution of the related precursor, up to the formation of the final material. The aim is to provide a scientific as well as practical basis for the selection of preparation methods for the production of materials with suitable morphological and phase composition properties.

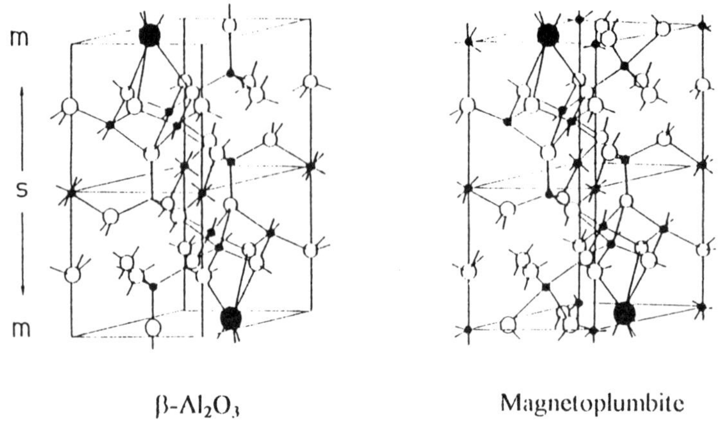

Figure 1 *Ideal semicell of* β-Al_2O_3 *and magnetoplumbite structures.*

Characterization studies of the final materials are also examined in order to clarify the relevant properties for high-temperature combustion, including: detailed structural features and their relation to sintering behavior and combustion activity; and role and nature of transition metal ion in the structure.

Finally, the use of hexaaluminates in gas turbine catalytic combustors is addressed.

2 Preparation Methods of Hexaaluminate Materials

Different methods are described in the literature to prepare hexaaluminate materials.

In Table 1 a survey of the preparation methods is reported with properties and target applications of these materials.

2.1 Solid State Reaction – Hexaaluminates have been long known as interesting materials for many applications (electrical ceramics, matrices for permanent immobilization of radioactive elements from nuclear wastes and refractory cement and concrete). For many years ceramists prepared hexaaluminates via solid state reaction for both crystallographic and application purposes.[6]

This method was adopted by Arai and co-workers[7-9] to prepare $MO{\cdot}6Al_2O_3$ systems (with M = Ca, Sr, Ba and Mg) in order to investigate their structural and morphological properties in catalytic applications. The starting reactants were finely divided powders of alkaline earth (Mg, Ca, Sr, and Ba) carbonates and γ-Al_2O_3 with mean crystal size of 800 nm and 5 nm, respectively. The powders were thoroughly mixed and ground in a mortar for 2 h before firing. Temperatures over 1400 °C were required to obtain the desired layered alumina phase for all the materials investigated except of MgO-Al_2O_3, which did not form this type

Table 1 *Preparation methods, temperature of formation, composition and morphology of the hexaaluminate systems*

System	*Method*	*Reagents*	*Temperature of formation of the final phase (°C)*	*Final phase*	*Applications*	*Ref.*
Ba,La-M-substituted hexaaluminate M = Mn	Solid state reaction	Oxides and/or carbonates and chlorides	1400-1600	Ba-β-Al_2O_3 or MP, no surface area data	Ceramic materials, Crystallography	6, 10
A-hexaaluminate A = Ba, Sr, Ca, Mg	Solid state reaction	Oxides and/or carbonates	1400-1600	Ba-β-Al_2O_3 or MP or spinels S_a=4 - 6 m^2/g	Catalytic combustion	7-9
Ba-Mn-substituted hexaaluminate	Solid state reaction with ball milling	Oxides and carbonates	1100-1300	Ba-β-Al_2O_3	Catalytic combustion	11
A′-M′-substituted hexaaluminate A′ = Ba, Sr, La M′ = Mn, Fe, Co, Ni	Sol-gel: alkoxides hydroyisis in alcoholic medium	A′ isopropoxides and M′ nitrates	1000-1200	Ba-β-Al_2O_3 or MP S_a=10-15 m^2/g	Catalytic combustion	5, 7-9, 15
Ba,La-hexaaluminate	Aerogel formation: alkoxides hydrolysis in alcoholic medium and supercritical drying.	Alkoxides	1300-1400	Ba-β-Al_2O_3 or MP extraphases always detected, S_a=10-15m^2/g	Catalytic combustion	18
Ba-hexaaluminate	Sol-gel: hydrolysis of organic precursors	Al alkoxy-ester and Ba acetate	1200-1300	Ba-β-Al_2O_3, no surface area	Ceramic meterials	19
La-hexaaluminate	Ammonia route: coprecipitation from nitrates aqueous solution	La and Al nitrates and ammonia solution	1000-1100	La-β-Al_2O_3, S_a=10-15 m^2/g	Catalytic combustion	21
A′-M′-substituted hexaaluminate A′ = Ba, Sr, La M′ = Mn, Fe, Co, Ni	Carbonates route: coprecipitation from nitrates aqueous solution	A′,M′-nitrates, ammonium carbonate	1000-1200	Ba,La-β-Al_2O_3 or MP, S_a=10-15 m^2/g	Catalytic combustion	22-24, 25, 27

of phase due to the small dimensions of the Mg^{2+} cations. The monophasic samples with layered alumina structure showed low surface area of 4.0-6.0 m^2/g.

Chandra[10] synthesized high purity hexaaluminate via solid state reaction at 1350 °C, starting from a powder mixture of $BaCO_3$ and $BaCl_2 \cdot 2H_2O$ (75/25 w/w) with aluminum oxide. A low melting $BaCO_3$-$BaCl_2$ eutectic formed at 840 °C that was reported to accelerate the reaction, thus favoring the formation of Ba-hexaaluminate at lower temperature than in the conventional method. However no surface area data were reported by this author.

Mechanochemical synthesis based on ball milling treatment of the powder mixture before calcination was also investigated.[11] A $BaMnAl_{11}O_{19}$ sample was prepared starting from different powder precursors such as oxides, hydroxides, and peroxides of the constituents. The powders were ball milled for 1-10 minutes before firing. A monophasic sample with Ba-β-Al_2O_3 structure was obtained at 1100 °C that retained a surface area of 8 m^2/g upon calcination at 1300 °C. Appropriate choice of the starting reactants and design of the mill were reported as the critical factors to achieve effective synthesis of these hexaaluminates.

2.2 Hydrolysis of Alkoxides – Although it is simple and economical, the solid state reaction route has the following major drawbacks for the preparation of catalysts: (1) very high temperatures are required to complete the formation of the final phase unless ball milling is performed before firing; (2) industrial catalysts often exhibit complex chemical composition that is difficult to obtain uniformly through solid state reaction; and (3) the low surface areas, that are typically achieved by solid state reaction, may affect the catalytic performances.

To obtain effective materials for catalytic combustion, Arai and co-workers developed a preparation method based on the hydrolysis of the alkoxides of the components.[12] This sol-gel method was originally developed to prepare high-purity submicron powders of alkaline or alkaline-earths titanates or zirconates.[13,14] The method is particularly suitable for low-temperature production of mixed oxides due to the effective mixing of the components at the atomic level that can be obtained in the gel precursor.

Ba-hexaaluminate samples were prepared by hydrolysis of commercial $Al(i\text{-}OC_3H_7)_3$ and $Ba(i\text{-}OC_3H_7)_2$ obtained from metallic Ba and 2-propanol in N_2 stream at 80 °C. The main steps of the preparation route are summarized in Figure 2. Stoichiometric amounts of the alkoxides were dissolved in 2-propanol and refluxed at 80 °C for 5 h under vigorous stirring in a dried N_2 atmosphere. Slow addition of water to the solution of the alkoxides caused gelation, accompanied by an immediate temperature rise from 80 to 90 °C. After ageing at 80 °C for several hours, evaporation to dryness in vacuum and calcination up to 1200 °C, the final Ba-hexaaluminate phase was obtained. The sample showed a surface area of 15 m^2/g that was retained up to 1450 °C. The hydrolysis ratio H_2O/ROH and the time of aging were investigated in details, because they markedly affect the properties of the final material.[5,8] It was shown that the hydrolysis process must be carefully completed before drying and calcination to obtain samples with high surface area. When the hydrolysis is not complete, pyrolysis of the unreacted alkoxides occurs during calcination, which results in

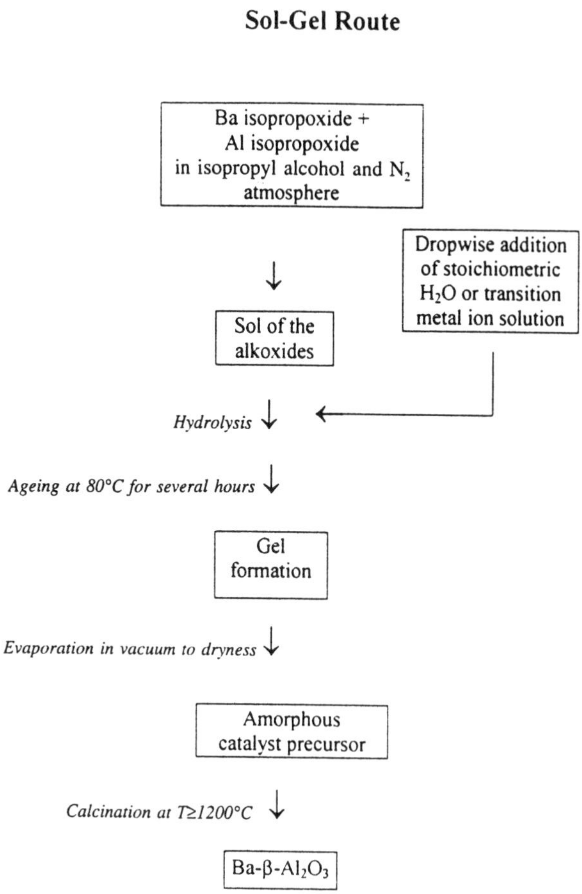

Figure 2 *Main steps of the preparation route via hydrolysis of alkoxides.*

the formation of large oxide particles. It was found that a stoichiometric amount of water ($H_2O/ROH=0.5$) must be used and 12 h of ageing are required to complete the hydrolysis process. The water excess must also be limited to prevent the formation of hydroxides in the precursor, which raised the temperature at which the final phase was formed and lowered the surface area.

This preparation method was successfully extended to the production of other hexaaluminates containing alkaline (K), alkaline-earth (Ca, Sr), and rare earth (La, Pr, Sm, Nd) metals in the mirror plane.[15,16] This route was also effectively used to prepare hexaaluminates incorporating transition metal ions in the structure (with M = Cr, Mn, Fe, Co, Ni, and Cu). In this case, nitrates of the transition metal ions were dissolved in the amount of water required for the hydrolysis.[5]

Because of its effectiveness and flexibility, this method was scaled up to the preparation of catalysts in the monolith form to be used in pilot and full-scale combustors.[17]

2.3 Other Sol-gel Processes – A sol-gel process similar to that proposed by Kyushu University was adopted by Lowe et al.[18] to prepare $LaAl_{11}O_{18}$ and $BaAl_{12}O_{19}$ samples. In this case, alcohol solutions of alkoxides and metal salts, such as Al and Ba isopropoxides and La nitrate, were used. Water in excess of stoichiometric amount needed for the hydrolysis reaction was added under vigorous stirring. The thick gel formed in less than 30 s and was thoroughly mixed and aged overnight. An aerogel precursor with very high surface area was obtained by supercritical drying at 300 °C. The aerogel was then calcined in the temperature range 1000-1600 °C. Results similar to those obtained by Arai were claimed, but according to the published data, phases other than layered alumina were always present.

To prepare hexaaluminates for ceramic applications a slightly different sol-gel process was proposed by Debsikbar.[19] Ba-hexaaluminates were prepared via hydrolysis of Al di(isopropoxide) acetoacetic ester chelate and anhydrous Ba acetate obtained by reaction between $BaCO_3$ and glacial acetic acid. The substitution of $Al(i\text{-}OC_3H_7)_3$ with the alkoxy ester was intended to control the chemical polymerization during gel formation. The reaction was performed in 1-butanol. The formation of the gel slowly occurred at room temperature in about 10 h. To obtain the final phase the gel precursor was dried at 70 °C for about 2 weeks, ground and calcined at 1200 °C for 2 h. However no data on the morphology of the final materials were reported by the author.

2.4 Co-precipitation in Aqueous Medium – Despite its effectiveness, the alkoxide method presents some drawbacks for industrial scale-up: the high cost of the starting reactants and the requirement for water-free and oxygen-free environment for the alkoxide solutions.

Most industrial catalysts based on mixed oxides are simply and economically prepared via co-precipitation in aqueous medium.[20] For the preparation of hexaaluminates, this method was first reported by S. Matsuda and co-workers.[21] $La_2O_3 \cdot xAl_2O_3$ samples were prepared starting from an aqueous solution of La and Al nitrates. The precipitation was carried out by the addition of NH_4OH solution up to pH=8. After it was washed, filtered and dried, the precursor was calcined at different temperatures up to 1400 °C. For a La_2O_3/Al_2O_3 mole ratio 5/95, the formation of a layered-alumina phase was observed starting from 1000 °C and samples with a surface areas of 30 m^2/g and 8 m^2/g have been obtained upon calcination at 1200 °C and 1400 °C for 2 h respectively.

The preparation of hexaaluminate catalyst by co-precipitation was systematically investigated by Groppi and co-workers.[22-24] In Figure 3, the main steps of the preparation procedure are sketched. A starting solution of the nitrates of the constituents was prepared by dissolving the precursor salts in water at 60 °C under vigorous stirring. The solution was acidified at pH=1 by HNO_3 to avoid precipitation of Al hydroxide at this stage. The nitrate solution was then poured

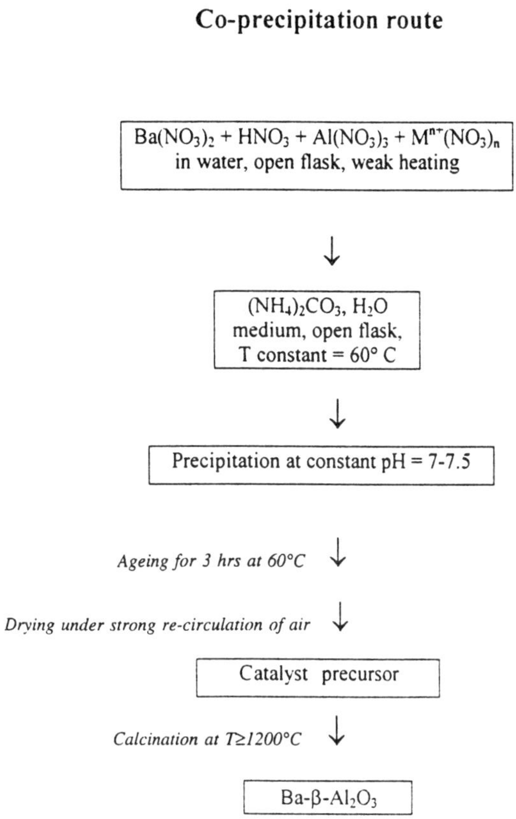

Figure 3 *Main steps of the preparation route via co-precipitation of carbonates.*

into an aqueous solution of $(NH_4)_2CO_3$ under vigorous stirring at constant temperature (60 °C). An excess of the precipitant agent $(NH_4)_2CO_3$ was used: (1) to favor the precipitation of alkaline-earth precursors; and (2) to self-regulate the pH through the buffer property of $(NH_4)_2CO_3$ solutions. Immediately after the nitrate and $(NH_4)_2CO_3$ solution were mixed the precipitation occurred, with evolution of carbon dioxide. The slurry was aged for 3 h at 60 °C. During precipitation and ageing the pH remained constant at 7.5-8.0. The precipitate was then filtered, washed for nitrate removal, and dried overnight at 110 °C under forced air circulation. Quantitative precipitation was confirmed by chemical analysis of the mother liquors and washing water. Monophasic samples with the layered-alumina structure were obtained at calcination temperatures between 1100 and 1300 °C, depending on chemical composition. Typical surface areas of 10-15 m^2/g were obtained upon calcination at 1300-1400 °C, which compare well with those provided by the alkoxide route.

This preparation method was successfully extended to a wide compositional range including M-substituted hexaaluminates (M = Mn, Fe, Cr, Co, and Ni,)[23-25] and Sr-La-based hexaaluminates.[26]

Jang et al.[27] adopted a similar method to prepare a BaMn-hexaaluminate and a LaSrMn-hexaaluminate series. These authors specifically investigated the alternative use of NH_4OH or $(NH_4)_2CO_3$ as precipitating agent. According to this study, the use of $(NH_4)_2CO_3$ provided materials with higher surface area than those obtained by using NH_4OH. Moreover, the hexaaluminate phase was formed in both precipitating agents, but monophasic samples were obtained using $(NH_4)_2CO_3$ only. Indeed, in the samples prepared with NH_4OH, other dominant phases such as α-Al_2O_3 were identified.

The effectiveness of the precipitation method based on the use of $(NH_4)_2CO_3$ in the preparation of $BaMn_1Al_{11}O_{19}$ was also confirmed by Berg et al.[28]

3 Thermal Evolution of the Precursor and Formation Mechanism of the Hexaaluminate Phases

The previous section describes different preparation methods, that produce materials with similar phase composition but markedly distinct morphological properties. To rationalize this behavior, the thermal evolution of the precursor phases obtained with the different preparation procedures has been thoroughly addressed in the literature.

The characterization data of $BaAl_{12}O_{19}$ samples obtained with different preparation procedures and of a pure alumina sample are summarized in Table 2, in which the phase compositions at relevant calcination temperatures are reported, and in Figure 4, in which the plots of surface area versus calcination temperature are compared.

3.1 Solid State Reaction – Starting from fine powders of $BaCO_3$ and γ-Al_2O_3 to give the final mole ratio BaO/Al_2O_3=0.14/0.86, Arai and co-workers[7-9] showed that the precursor underwent a complex formation process to give the final Ba-β-Al_2O_3 phase. Upon calcination at 1000 °C, the sample consisted of a mixture of γ-Al_2O_3 and well crystallized $BaAl_2O_4$ that was reported to form by a relatively easy solid state reaction between $BaCO_3$ and γ-Al_2O_3. The formation of Ba-hexaaluminate phase started at 1200 °C and was completed only upon calcination at 1450 °C. At this temperature, a monophasic material was obtained, that was stable up to 1600 °C. Above 1100 °C a progressive drop of surface area was observed that paralleled the formation of the Ba-hexaaluminate phase. The surface area stabilized at about 6 m^2/g above 1300 °C. Microstructural analysis by SEM micrographs[8] showed that particles of the hexaaluminate phase were strongly agglomerated with size of the order of 500-1000 nm, after calcination at 1450 °C.

The following mechanism for the formation of the final phase was proposed. At relatively low temperature, $BaCO_3$ and γ-Al_2O_3 particles react to form the nonequilibrium phase $BaAl_2O_4$. The formation of the stable Ba-hexaaluminate phase starts at 1200 °C and occurs via a very slow solid state reaction between

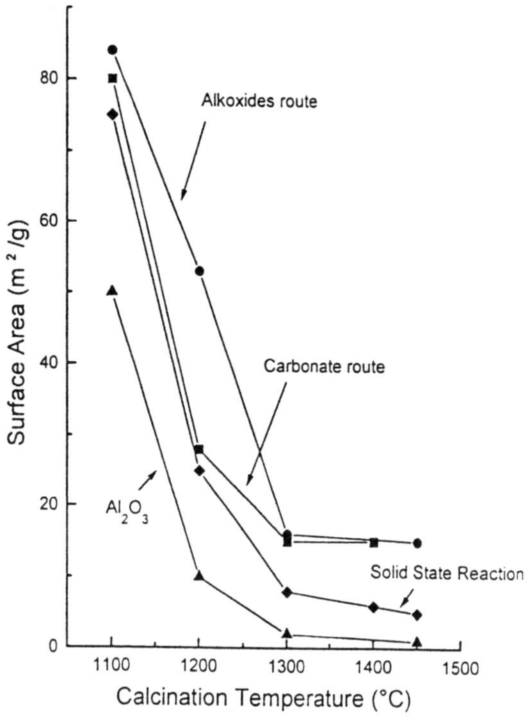

Figure 4 *Surface area vs. calcination temperature of $BaAl_{12}O_{19}$ obtained with different preparation route and of pure Al_2O_3.*

well crystallized $BaAl_2O_4$ and Al_2O_3 particles. At this temperature, other competing processes are likely to occur. Indeed, as shown by the surface area data of pure alumina samples reported in Figure 4, a marked sintering of the Al_2O_3 particles is evident, usually accompanied by the formation of α-Al_2O_3. In the case of Ba-Al-O sample, this latter phenomenon is prevented by the formation of Ba-β-Al_2O_3, but sintering still occurs. This competing sintering process along with the presence of well crystallized $BaAl_2O_4$ results in the formation of Ba-β-Al_2O_3 particles with relatively large crystal size.

3.2 Hydrolysis of Alkoxides – A different thermal evolution of the precursor phase was observed for a $BaAl_{12}O_{19}$ sample prepared via hydrolysis of alkoxides.[8,9] For this sample no XRD diffraction lines were detected after heating at 1000 °C. At 1200 °C Ba-hexaaluminate formed, and it was the only phase detected in the XRD spectra. Calcination at 1450 °C resulted in higher sharpness and intensity of the XRD reflection of Ba-hexaaluminate. No further changes were observed upon calcination at 1600 °C. Also in this case, the appearance of Ba-β-Al_2O_3 was accompanied by a marked drop of surface area. However, the stable value of surface area obtained for calcination temperature above 1300 °C (15 m^2/g) was 3 times larger than that obtained via the solid state reaction route. The superior morphological properties of the sample prepared via hydrolysis of alkoxides were confirmed by SEM analysis that showed the presence of smaller crystal grains than those obtained by solid state reaction. No intermediate phases, such as $BaAl_2O_4$, were observed. Accordingly, the Ba-hexaaluminate phase directly formed from the thermally decomposed gel precursor. Arai and co-workers suggested that the intimate contact between reacting species provided in the amorphous precursor favors the formation of the final layered-alumina phase at lower temperatures than those required by solid state reaction. The easier formation of Ba-β-Al_2O_3, along with the absence of sintered precursor phases, would cause the superior morphological properties obtained with the alkoxide method.

3.3 Co-precipitation in Aqueous Medium – The co-precipitation route can also in principle allow for high interspersion of the constituents in the precursor. In the case of $BaAl_{12}O_{19}$, the very small dimensions of the aggregates obtained with the specific conditions adopted for the precipitation was evidenced by the high surface area (250 m^2/g) measured for the dried sample. However, at variance with the alkoxide route, a complex phase evolution is observed on increasing the calcination temperature.[22] Starting from a microcristalline precursor consisting of γ-Al_2O_3 and dispersed Ba species, at 900 °C well crystallized $BaAl_2O_4$ formed as intermediate product. XRD quantitative analysis showed that the amount of Ba segregated as $BaAl_2O_4$ increases up to 40% of the total Ba at 1000 °C, then progressively decreased and completely disappeared at 1300 °C. The remaining fraction of Ba was detected as amorphous carbonate species whereas excess alumina was present as γ-Al_2O_3 up to 1200 °C, i.e., well beyond the transition temperature observed for the pure alumina sample (see Table 2).

Table 2 *Phase evolution of the Ba-Al-O system prepared via different preparation methods*

System	*Phase composition*			
	T_c = 1000 °C	*T_c = 1200 °C*	*T_c = 1450 °C*	*T_c = 1600 °C*
Al_2O_3	θ-Al_2O_3	α-Al_2O_3	α-Al_2O_3	α-Al_2O_3
Ba-hexaaluminate via solid state reaction	$BaAl_2O_4$ + γ-Al_2O_3	Ba-β-Al_2O_3 + $BaAl_2O_4$	Ba-β-Al_2O_3	Ba-β-Al_2O_3
Ba-hexaaluminate via alkoxides route	Amorphous	Ba-β-Al_2O_3	Ba-β-Al_2O_3	Ba-β-Al_2O_3
Ba-hexaaluminate via carbonates route	$BaAl_2O_4$ + γ-Al_2O_3	Ba-β-Al_2O_3 + traces of $BaAl_2O_4$	Ba-β-Al_2O_3	Ba-β-Al_2O_3

XRD reflections of the Ba-β-Al_2O_3 phase were detected in the spectra starting from 1100 °C and their intensities progressively increased up to 1300 °C. At this temperature Ba-β-Al_2O_3 was the only phase detected by XRD. No further phase evolution was observed up to 1600 °C. Also in this case a marked drop of surface area paralleled the formation of the layered-alumina phase. However, despite the presence of crystalline $BaAl_2O_4$ as reaction intermediate, a stable surface area of 15 m^2/g was retained above 1300 °C, associated with particles with nearly hexagonal contour 100-500 nm wide.[29]

Considering that the structure of the Ba-β-Al_2O_3 phase is closely related to that of the precursor γ-Al_2O_3 matrix, (because it consists of alternately stacked γ-Al_2O_3-type spinel blocks and Ba-containing mirror planes), the following formation mechanism is suggested. Starting from a threshold temperature of 1100 °C Ba-β-Al_2O_3 forms via solid state reaction between microcrystalline γ-Al_2O_3 and dispersed Ba compounds or crystalline $BaAl_2O_4$. This reaction proceeds through the diffusion of Ba ions, along the oxygen close-packed planes, in the spinel structure of γ-Al_2O_3. For energetic reasons, the presence of Ba in adjacent crystallographic planes is prevented, thus the diffusion process results in the typical layered structure of Ba-β-Al_2O_3. According to this mechanism the observed threshold temperature of 1100 °C is related to the activation of the diffusion of large Ba ions within the alumina matrix.

The key factor in achieving the final hexaaluminate material with high surface area is likely related to the stabilization of γ-Al_2O_3 up to the formation temperature of Ba-β-Al_2O_3. This is caused by the presence of the dispersed Ba species provided by the coprecipitation route. The literature[30,31] shows that addition of small amounts of alkaline-earth and rare-earth metal oxides onto the alumina surface can effectively inhibit sintering and phase-transition phenomena. The stabilization of the alumina matrix has two positive effects on the formation process: (1) the presence of small aggregates markedly favors solid state reactions; and (2) the inhibition of the competing sintering process prevents irreversible depletion of surface area.

The solid state reaction route, starting from large $BaCO_3$ particles, cannot provide the required dispersion of Ba. Accordingly, the growth of the alumina particles is not inhibited and, because of this, the formation of the hexaaluminate is more difficult and requires very high temperature. Also, the morphological properties are negatively affected.

The characterization data obtained for the carbonate route prove that the intermediate formation of well crystallized $BaAl_2O_4$ does not prevent the production of materials with high surface area.

3.4 Effect of the Transition Metal Ions – Hexaaluminate materials, $BaMAl_{11}O_{19}$, including transition metal ions in the structure (M = Mn, Fe, Cr, Co, Ni) were prepared both via the alkoxide[15] and the coprecipitation route.[23,24,25] For all the compositions investigated, monophasic samples with layered-alumina structure and surface area in the range 10-15 m^2/g were obtained upon calcination at 1300 °C.

Groppi and co-workers[23,24] investigated the thermal evolution of Mn- and Fe-substituted hexaaluminates with different amounts of transition metal ions prepared via the co-precipation route. The dried precursors of $BaMn_xAl_{12-x}O_{19}$

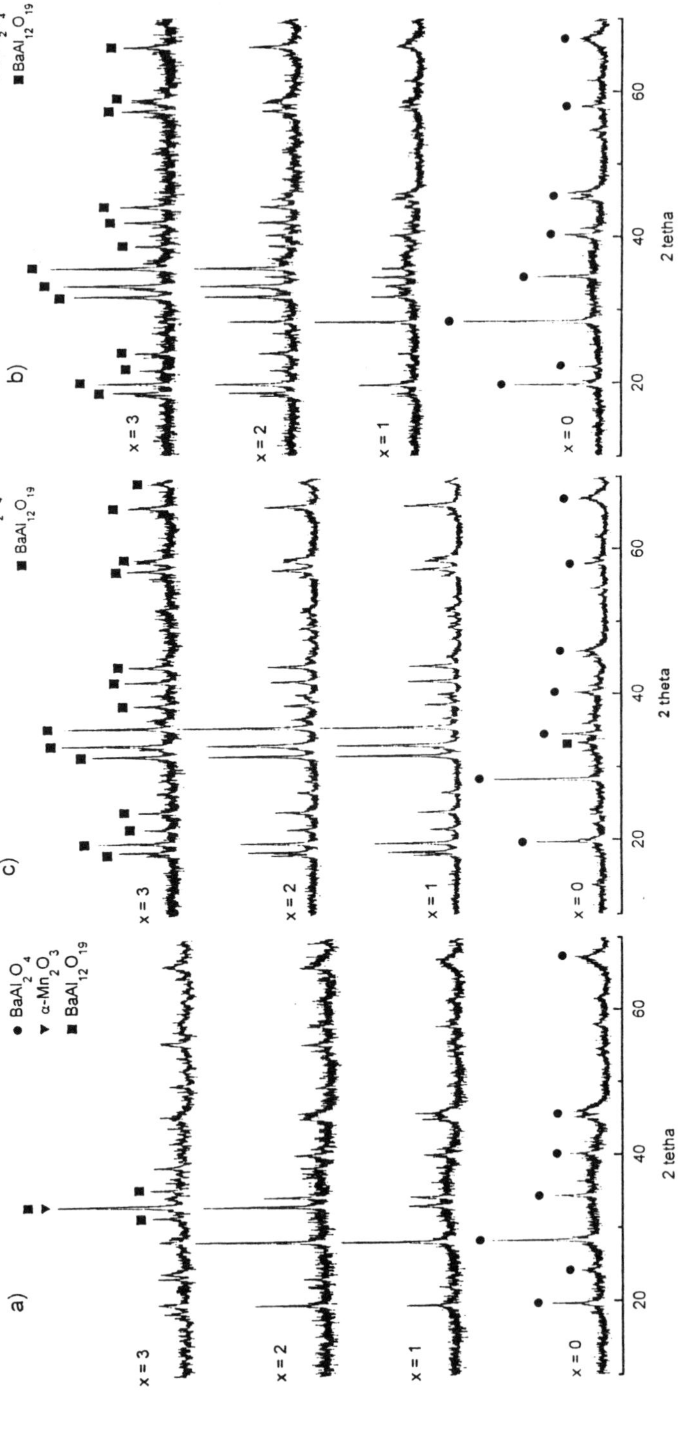

Figure 5 *XRD spectra of $BaMn_xAl_{12-x}O_{19}$ with x= 0, 1, 2, 3.*
(a) calcination temperature 900 °C; (b) calcination temperature 1000 °C; (c) calcination temperature 1100 °C.

with x=0.5, 1, 2, 3, showed similar composition to that of unsubstituted $BaAl_{12}O_{19}$. They consisted of an amorphous matrix with high surface area (200-250 m^2/g), including metastable crystalline phases that easily decompose upon calcination at 500 °C. After heating in air at 700 °C, all the samples consisted of microcrystalline γ-Al_2O_3, crystalline α-Mn_2O_3, (whose amount increased with Mn loading), and dispersed Ba species. In Figure 5a-c, the XRD spectra of $BaMn_xAl_{12-x}O_{19}$ with x= 1 2, 3 calcined at 900, 1000 and 1100 °C, are compared with those of a $BaAl_{12}O_{19}$ sample. A similar phase composition was observed for all the samples, consisting of $BaAl_2O_4$, α-Mn_2O_3 and Ba-β-Al_2O_3. However, a promoting effect of Mn on the formation of the final layered-alumina phase was evident.[23] At 900 °C, the reflections of Ba-β-Al_2O_3 were present only in $BaMn_3Al_9O_{19}$, i.e., the sample with maximum Mn loading. Upon calcination at 1000 °C the formation of this phase started in all the Mn-containing sample; the relative amount of Ba-β-Al_2O_3 increases with the Mn-loading. Only the reflections of well crystallized $BaAl_2O_4$ were present in the spectrum of $BaAl_{12}O_{19}$. At 1100 °C, i.e., the threshold temperature for the formation of Ba-β-Al_2O_3 in $BaAl_{12}O_{19}$, the reaction to the layered-alumina phase was completed in all the Mn-containing samples. They showed monophasic composition except for $BaMn_3Al_9O_{19}$; this latter sample segregated small amounts of $BaAl_2O_4$ (2% w/w), as revealed by Rietveld analysis of very carefully collected XRD powder spectra, and $BaAl_2O_4$ was still present upon calcination at 1300 °C.

The same formation mechanism of Ba-β-Al_2O_3 proposed for unsubstituted material was hypothesized for Mn-substituted hexaaluminates. The promoting effect of Mn was related to an increase of ion mobility within the γ-Al_2O_3 spinel blocks that, according to the above mechanism, would result in the lowered threshold temperature for the Ba-β-Al_2O_3 formation. The increasing ion mobility within the γ-Al_2O_3 lattice due to the presence of Mn, was supported by the comparison of the temperatures of phase transition of pure and Mn-doped alumina samples. It was shown that the addition of Mn resulted in a lower transition temperature. Because the alumina transitions imply lattice rearrangements that require mobility of oxygen ions, these data provide evidence for a promoting effect of Mn on ion mobility.[25]

Also in Fe-containing materials, promoting effect on the formation of the final layered-alumina phase was observed.[24] In the completely substituted material $BaFe_{12}O_{19}$ a monophasic sample with magnetoplumbite structure was obtained at 700 °C. This low formation temperature was related to the greater mobility of oxygen and Ba ions in the lattice of Fe oxides than in Al oxides. Indeed, the transition from $\gamma \rightarrow \alpha$ alumina occurs in Fe oxides at temperatures hundreds of degrees lower than those required for phase transitions in Al oxides.

4 Characterization of Hexaaluminate Properties Relevant to Catalytic Combustion

Materials for high-temperature catalytic combustion should possess both elevated activity and high thermal stability. Generally these are opposite features,[32,33] so that a compromise should be provided by catalysts suitable for practical applications. Hexaaluminates have the potential to meet the requirements of industrial combustor. Their peculiar crystal structure can provide the stable morphology and phase composition necessary to ensure stable performance under the severe operating conditions of the combustor and can allow for incorporation of transition metal ions that are responsible for combustion activity.

The relatively high surface area of hexaaluminates upon calcination at high temperature is relevant to catalytic performance. Although the role of catalyst surface area is a debated issue in the technical and scientific literature on catalytic combustion for gas turbine applications, it is widely accepted that, for effective operation of the monolith combustor, heterogeneous reactions must begin immediately after the monolith inlet. Thus, almost the whole catalyst bed operates in the diffusion-controlled regime. In such conditions only the external skin of the catalyst participates in combustion because the reaction rate is too fast to allow for reactant diffusion within the pores. Accordingly, some authors have proposed using an active material with very low but stable surface area, thus avoiding the high temperature sintering, that is one of the main problems to be faced in catalyst development.[34] On the other hand, high catalyst activity is usually obtained by combining a high specific activity, i.e., high turnover frequency, with many active sites, which is directly related to the catalyst surface area. Mathematical analysis of combustor transients showed that larger surface area would be beneficial for catalyst ignition during the start up.[35] Moreover, simulations of ceramic catalysts with negligible heat conduction of the solid wall showed that a certain amount of surface area (i.e., bulk activity) is required to stabilize the catalyst ignition at steady state by balancing the quenching effect of the incoming cold gas stream and the heat production in the solid.[36]

The structural features of hexaaluminates, their sintering behavior, and their catalytic properties associated with the nature of the transition metal ion are discussed in detail in the following section.

4.1 Structural Properties and Sintering Behavior – The sintering resistance of hexaaluminates is strictly related to the unique layered structure of these materials.[7-9] As mentioned above, the layered hexaaluminate structures can be classified into two types: magnetoplumbite and β-Al_2O_3. They both consist of closely packed spinel blocks separated by mirror planes that contain large cations of alkaline, alkaline earth and rare-earth ions and loosely packed oxygen ions. Spinel blocks and mirror planes are alternatively stacked, originating an hexagonal structure with the c axis parallel to the direction of stacking (Figure 1).

Transmission electron microscopy (TEM) analysis reveals that these materials crystallize as hexagonal planar particles with marked anisotropic shape,[8,37] as shown in Figure 6. When appropriate preparation methods are used, plate-like crystals are obtained with small thickness of about 20-30 nm and an aspect ratio D/h=5-10. Selected area diffraction (SAD) patterns of incident beams perpendicular and parallel to the large hexagonal facet show that they correspond to the crystallographic planes perpendicular to the c axis. The anisotropic shape of the

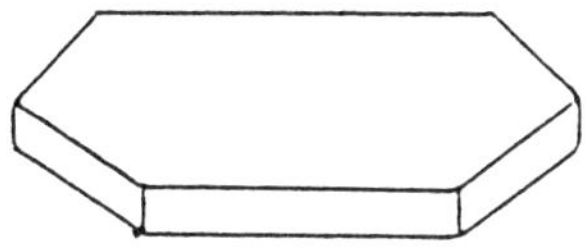

Figure 6 *Sketch of the thin planar crystallites of* $BaAl_{12}O_{19}$.

crystallites, shows that a strong suppression of crystal growth along the c axis occurs, causing preferential grain growth along the direction perpendicular to c. However, because particles with a very large aspect ratio are unstable (because of high surface energy), the average particle size does not increase easily at elevated temperature; this stabilizes surface area.

Diffusion of ions in the solid lattice is generally recognized as the governing factor in sintering mechanism.[32] Secondary-ion mass spectrometry (SIMS) analyses of ^{18}O ions[38,39] performed over a single crystal of Mn-substituted Ba-hexaaluminate with Ba-β-Al_2O_3 structure showed that O^{2-} diffusion along the c axis is one order of magnitude lower than in the orthogonal direction. In the loosely packed mirror planes of layered-aluminas, relatively fast diffusion is expected due to the presence of intermediate available sites. Indeed, materials with β-Al_2O_3 structure are well known conductive ceramics due to high mobility of large cations (e.g., Na^+) in the mirror planes.[6] On the other hand, the resistance to O^{2-} diffusion along the c axis is associated with the difficulty of large ions to migrate into the cubic close packing of noncationic deficient spinel blocks. The inhibition of ion diffusion along the stacking direction is probably responsible for suppression of crystal growth along the c axis.

A detailed investigation of the crystal structure of Ba-Al-O system performed by Rietveld analysis of powder diffraction spectra[40] provided further information on the mechanism of sintering. In line with previous studies on single crystal materials,[41-43] monophasic samples with a layered-alumina phase were obtained in the compositional range from Al/Ba=14.67 to Al/Ba=9.15. Two different phases were observed: β_I of composition $Ba_{1.5}Al_{22}O_{34.5}$ and β_{II} of composition $Ba_{2.33}Al_{21.33}O_{34}$. The structure of both phases contains defective and nondefective cells. In the case of β_I the nondefective cell of formula $[Ba_2O_2Al_{22}O_{32}]^{2+}$ consists of two spinel blocks $[Al_{11}O_{16}]^+$ separated by two mirror planes of composition [BaO], while the defective cell of formula $[O_4Al_{22}O_{32}]^{6-}$, originates from the presence of Ba vacancies inducing complex defect structures. To maintain electroneutrality, three nondefective cells are present for each defective one, resulting in the above theoretical composition. The β_{II} phase also consists of the same nondefective cell of β_I and of defective cells of composition $[Ba_3Al_{20}O_{35}]^{4-}$ where Ba ions are allocated also in the spinel blocks. Electroneutrality requires two nondefective cells for each defective one.

For composition near the extremes of the formation range of layered-alumina phases, samples with typical β_I (Ba poor composition) and β_{II} (Ba-rich composition) structure were obtained. For the intermediate composition, $BaAl_{12}O_{19}$,

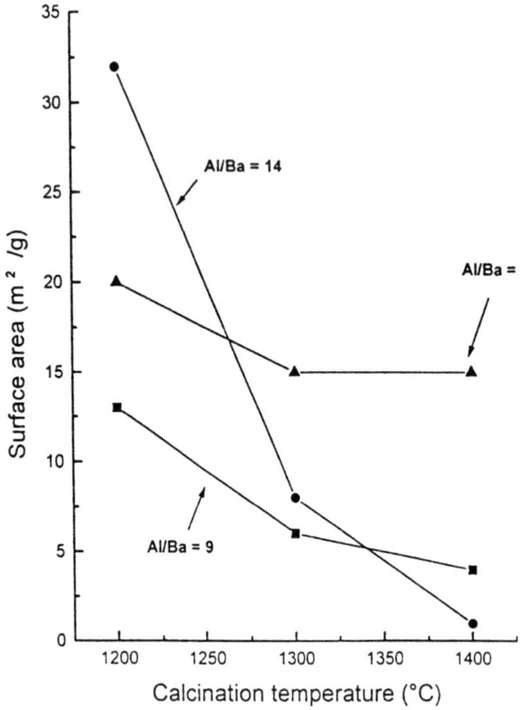

Figure 7 *Surface area of the Ba-Al-O samples with Al/Ba= 9, 12 and 14 vs. calcination temperature.*

a phase characterized by a structure intermediate between β_I and β_{II} was observed. This phase likely originates from the intergrowth of β_I and β_{II} domains permitted by the close similarity of the two structures.

In Figure 7, the surface area of the three samples with Al/Ba= 9, 12, and 14 are reported as a function of calcination temperature in the 1200-1400 °C range. BET measurements indicated that high surface area and sintering resistance (15 m^2/g at 1400 °C) were obtained only for the intermediate composition $BaAl_{12}O_{19}$. The Ba-poor sample showed high surface area at 1200 °C but sintered markedly at higher temperatures even though the formation of Ba-β_I-Al_2O_3 was complete at 1200 °C. On the other hand, the formation of Ba-β_{II}-Al_2O_3 phase in the Ba-rich sample was complete only at 1300 °C, likely due to the difficult Ba allocation within the spinel blocks. Thus, irreversible sintering of the precursor phase occurred that resulted in low surface area of the final material.

These data indicate that sintering resistance is related not only to the formation of a layered-alumina structure. It appears that a critical Ba content is required for the effective suppression of crystal growth along the c axis of crystallites. Indeed, in $BaAl_{14}O_{22}$, sintering proceeded even after the formation of Ba-β-Al_2O_3.

Further characterization of $BaAl_{12}O_{19}$ based on FTIR of adsorbed probe molecules (CO_2 and pyridine) and measurements of isoelectric point[29] showed that the surface was basic. Because of the low bulk Ba content of this material such basic character can hardly be explained assuming a statistical exposure of the different crystallographic planes. This suggests that the Ba-containing mirror planes are preferentially exposed.

In conclusion, the sintering resistance arises from the difficulty of crystallites with Ba-containing planes exposed at the surface to coalesce along the c axis due to the following reasons: (1) the allocation of adjacent Ba-containing planes is not energetically permitted; and (2) the resistance to Ba diffusion along the c axis prevents the rearrangement of the stacking sequence.

Most of the studies reported above were performed using unsubstituted hexaaluminates. The addition of transition metal ions in the structure was reported to affect the sintering behavior at different extents depending on the concentration and the nature of the metal. In $BaMn_xAl_{12-x}O_{19}$ calcined at 1300 °C the incorporation of Mn had a minor effect on surface area up to x=1, whereas at higher Mn contents, the surface area gradually decreased with Mn loading.[5,23,44] The promotion of sintering at high Mn loading was associated with the increased diffusion rate of Ba ions in Mn-containing spinel blocks, which was also responsible for more rapid formation of the final layered-alumina phase. This was confirmed by the calculated aspect ratio D/h of the plate-like crystallites, which decreases with increasing Mn content. The absence of a significant sintering effect in $BaMn_xAl_{12-x}O_{19}$ with $x \leq 1$ could be associated with the peculiar location and nature of Mn ions in these samples (which will be addressed in the following sections).

The incorporation of Fe ions in layered-alumina structure was reported to favor sintering as well.[24] In particular, the completely substituted $BaFe_{12}O_{19}$ sample with the magnetoplumbite structure markedly sintered upon calcination above 1000 °C. Also in this case, a strong analogy was suggested between the promoting effect of Fe on the formation of the final phase and on sintering; indeed both processes are related to the greater mobility of Ba in the spinel blocks of Fe oxides compared to Al oxides. Mixed $BaFe_xAl_{12-x}O_{19}$ samples show intermediate behavior between $BaAl_{12}O_{19}$ and $BaFe_{12}O_{19}$ with sintering resistance that gradually increases with the Al content.

Arai and co-workers investigated the effects of cation composition in the mirror plane of Mn-substituted hexaaluminates.[5,15] Investigation of a series of rare-earth-based hexaaluminates ($AMnAl_{11}O_{19}$, A=La, Pr, Nd, Sm) prepared via hydrolysis of alkoxides and calcined at 1200 °C for 5 h showed that surface area increases with the ionic radius of A. The La-substituted sample, the largest cation of the series, showed a surface area of 15 m^2/g, which compared with 5 m^2/g for the Sm-substituted one, i.e., the smallest cation of the series. These data are consistent with the mechanism of sintering resistance reported above. Indeed, it is expected that the larger the cation, the more difficult its diffusion along the c axis.

Partial replacement of cation in the mirror plane was also studied for $AA'Mn_1Al_{11}O_{19}$ samples (A=Ba, Sr, Ca, La, K; A'= Sr, Ca, La, K). It was reported that substitution of cations in the mirror plane resulted in enhanced surface area only when aliovalent ions were used with size close to that of the host cation. The highest surface areas upon calcination at 1300 °C were obtained for the following compositions: $Sr_{0.8}La_{0.2}Mn_1Al_{11}O_{19}$ (24 m^2/g) and $Ba_{0.8}K_{0.2}Mn_1Al_{11}O_{19}$ (23 m^2/g). The former composition was selected for the industrial scale up of hexaaluminate catalysts.[17]

This surface area enhancement was not observed when preparing material with the same composition according to the co-precipitation route.[26,27]

4.2 Catalytic Combustion Properties of M-substituted Hexaaluminates – Most of the catalytic studies performed over hexaaluminate materials deal with the combustion of CH_4 as the main component of natural gas, i.e., the typical fuel of gas turbines. Arai and co-workers were the first to investigate the CH_4 combustion activity of $BaMAl_{11}O_{19}$ with M=Cr, Mn, Fe, Co, Ni prepared via the alkoxide route.[5] Activity tests were performed over powder catalysts using a conventional quartz microreactor fed with a diluted CH_4-air mixture (1% CH_4) at high-space velocity (GHSV=48000 h^{-1}). The results are summarized in Table 3 in terms of $T_{10\%}$ (i.e., the temperature required to achieve 10% conversion).

Table 3 *Methane combustion activities of $BaM_1Al_{11}O_{19}$*[5]

M	$T_{10\%}$ (°C)
Al	710
Cr	700
Mn	540
Fe	560
Co	690
Ni	710

Unsubstituted $BaAl_{12}O_{19}$ sample showed very poor activity. For the samples containing transition metal ions, the following order of reactivity was found Mn≅Cu>Fe>>Co≅Cr≅Ni. The same reactivity order was found over $LaMAl_{11}O_{19}$ samples,[15] which indicated that combustion activity is primarily determined by the nature of the transition metal ion. As shown in Figure 8, a volcano correlation was found between catalytic activity (expressed as $T_{10\%}$) and the difference of standard formation enthalpies of trivalent and divalent M-oxides. This suggests that CH_4 oxidation occurs according to a redox mechanism that involves a reversible $M^{3+} \leftrightarrow M^{2+}$ cycle. The volcano relation indicates that the reduction of M^{3+} into M^{2+} is the slowest step in CH_4 oxidation over the catalysts located in the right side, whereas $M^{2+} \rightarrow M^{3+}$ reoxidation is the rate-determining step over the transition metal ion to the left side of Figure 8. The most active catalysts originate from an appropriate balance between rates of reduction and reoxidation steps.

La-hexaaluminates showed higher catalytic activity for all the M-substituted systems with respect to Ba-hexaaluminates.[15,27] A similar promoting effect was also found for the composition $Sr_{0.8}La_{0.2}Mn_1Al_{11}O_{19}$,[15,16,26] which also exhibits the maximum surface area, but only when prepared via hydrolysis of alkoxides.

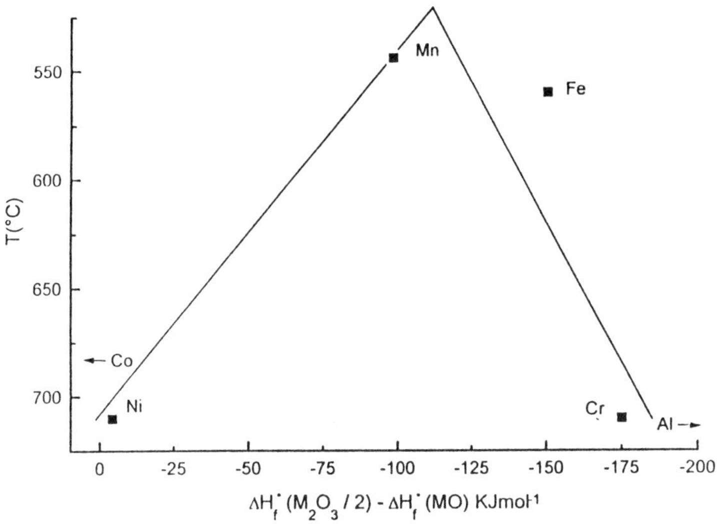

Figure 8 *Volcano correlation between catalytic activity and the difference of standard formation enthalpies of trivalent and divalent M-oxides.*

However, the extent of the activity enhancement cannot be related to the higher surface area of this material. Two possible explanations were proposed to account for the effect of mirror plane composition on combustion activity: one is related to the different oxidation state of the cation in the mirror plane; the other is associated with the crystal structure of layered-alumina materials (i.e., magnetoplumbite and β-Al_2O_3) which have different population and co-ordination of the ions in the mirror planes. Both these electronic and structural factors can, in principle, affect the redox properties.

The effect of the concentration of transition metal ion in the structure was also addressed. Specifically, the methane combustion activity of $BaMn_xAl_{12-x}O_{19}$[5,23,44] and $BaFe_xAl_{12-x}O_{19}$[24] was investigated. The catalytic activity of Mn-substituted systems was found to increase up to x=2 (Figure 9). No variations were observed on increasing the Mn content at x=3 whereas further Mn substitution resulted in a marked activity drop. Such behavior was associated with the limited capability of Ba-β-Al_2O_3 to incorporate Mn in the structure. Indeed, for Mn content $x \geqslant 3$, segregation of extra-phases was observed; moreover, a marked decrease of surface area occurred for x>1.

No limits on the content of transition metal ion were found for Fe-substituted systems. As mentioned above, and in line with crystallographic literature,[45] a monophasic sample with magnetoplumbite structure was obtained for the completely substituted $BaFe_{12}O_{19}$. This sample showed excellent CH_4 combustion activity upon calcination at 700 °C, but it markedly deactivated either upon increasing the calcination temperature or upon treatment at high tempera-

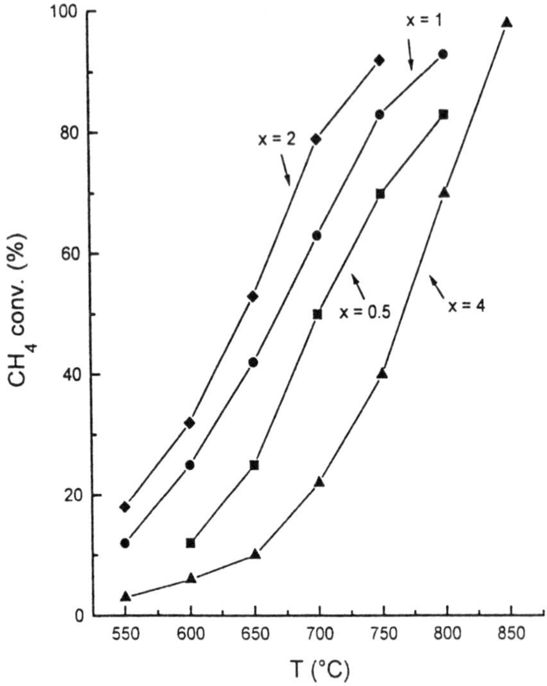

Figure 9 *Catalytic activity of $BaMn_xAl_{12-x}O_{19}$ (x = 0.5, 1, 2 and 4) vs. reaction temperature.*

ture under reaction conditions. Accordingly, this system is not useful for practical applications at high temperature.

Groppi and co-workers[46,26] investigated the catalytic activity of Mn- and Fe-substituted hexaaluminates prepared via the carbonate route for CO and H_2 combustion. These species are the main components of fuels from gasification of carbon and biomasses that represent an alternative to natural gas in gas turbine applications.

In line with general indications on the reactivity order over both mixed oxide and noble metal catalysts, CO and H_2 were found to be much more reactive than CH_4 over all the investigated hexaaluminate catalysts. Specifically, the following reactivity order was determined over Mn-substituted samples: $CO>H_2>>CH_4$. $T_{10\%}$ values of 230 °C and 320 °C were obtained over $BaMnAl_{11}O_{19}$ for CO and H_2, respectively, to be compared with 540 °C required by CH_4 combustion under similar experimental conditions. Apparent activation energies for CO and H_2 combustion were calculated to be 10 kcal/mol lower than that of CH_4 combustion (13-15 kcal/mol vs. 21-23 kcal mol), in line with the marked activity differences.

$Sr_{0.8}La_{0.2}Mn_1Al_{11}O_{19}$ was found to be more active than $BaMnAl_{11}O_{19}$ also in

CO and H_2 combustion although the activity enhancement was less marked than in CH_4 combustion. Again Fe-substituted systems were not stable under reaction conditions.

All the activity data discussed above were collected over powder catalysts in a lab-scale microreactor. The operating conditions in real combustors are characterized by much greater space velocities, pressures, and temperatures. Transport of heat, mass, and momentum have strong effects on the combustor performances. Detailed mathematical models of the monolith combustor were used for extrapolation of the lab-scale data.[36] In the case of CH_4 combustion, simulation results pointed out that, under real conditions, ignition occurs only above 750-800 °C. Accordingly, hexaaluminate catalysts can be used only as stable end-stage catalysts in segmented configurations in which much more active catalysts are used as low temperature igniters in the first catalyst stages. In the case of CO/H_2-containing mixtures, the ignition temperature over hexaaluminates is roughly 300 °C lower, but it is still not enough low to fulfill the requirements of gas turbine combustor in the most demanding conditions (idle and partial load).

4.3 Nature and Role of the Transition Metal Ion in the Structure – Catalytic combustion over M-substituted hexaaluminates occurs via a redox mechanism involving reversible variation of the oxidation state of the transition metal ion in the structure. Because of this, the nature and the role of the transition metal ion in the structure have been extensively investigated.

The average oxidation states of the transition metal ions in $BaMAl_{11}O_{19}$ catalysts were determined by thermogravimetry.[5] The samples calcined at 1300 °C were pre-treated in-situ at 1000 °C in dry air, cooled at room temperature and then heated up to 1100 °C in an H_2 stream. The degree of reduction was determined by the weight loss of the sample. By proper assumption of the final oxidation state, the results reported in Table 4 were obtained. Average oxidation states ranging from 2.0 to 3.0 were found depending on M. The values correlate well with the M-O bond strength, which is estimated as the standard heat of formation of the oxides. The higher the M-O bond strength, the higher the oxidation state. In line with the volcano correlation previously described, both the transition metal ions with low M-O strength (Ni, Co) that are present in the divalent state, and those with high M-O strength (Cr, Al) that are present in the trivalent state, show poor combustion activity. Intermediate oxidation state between 2+ and 3+ was found for the most active element: Mn.

Table 4 *Average oxidation state of M in $BaMAl_{11}O_{19}$*[5]

M	*Average oxidation*
Cr	3.0
Mn	2.4
Fe	3.0
Co	2.1
Ni	2.0

Temperature Programmed Desorption (TPD) of oxygen was also performed over $BaMAl_{11}O_{19}$ samples.[5] Very small amounts of oxygen desorbed from the poorly active Cr-, Ni- and Co- substituted catalysts whereas large amounts of oxygen were desorbed at high temperature (>500 °C) from the active Mn- and Fe- substituted systems. Oxygen desorption was associated with the reduction of the transition metal ion in the hexaaluminate lattice since no desorption peaks were measured without metal substitution. Mn- and Fe-substituted catalysts also showed an exothermic peak when heated in air after TPD measurements, indicating that the re-oxidation of the reduced transition metal ion can easily occur. These results confirmed that catalytic activity is related to reversible redox cycle of the transition metal ion in the structure.

The nature of the transition metal ion in the structure was further investigated by means of spectroscopic techniques. Fundamental studies on single crystals provided evidence for the presence of divalent Mn in Al(2) tetrahedral sites of both β-Al_2O_3 and magnetoplumbite-substituted materials.[39,47]

Bellotto and co-workers[44] investigated the Mn location and valence at the different sites in polycrystalline substituted Ba-β-Al_2O_3 with Mn/Ba ratio varying from 0.5 to 3.0. Rietveld refinements of multiple diffraction powder data sets, collected at different wave lengths in proximity to, and far from, the Mn-K-absorption edge with both synchrotron and Cu-Kα radiation, were used for this purpose along with X-ray absorption near edge spectroscopy (XANES). This study showed that at low concentration Mn preferentially enters the structure in tetrahedral Al(2) sites with dominant oxidation state 2+ (Figure 10). Replacement of Al^{3+} by Mn^{2+} is associated with the reduction of Ba vacancies in the mirror plane, and this mechanism of charge compensation allows insertion of an appreciable amount of Mn in the Ba-β_I-Al_2O_3 structure without phase separation.

On increasing the Mn content, Ba sites are progressively saturated and defects in the mirror plane disappear. Accordingly, the charge compensation mechanism can no longer operate, and Mn preferentially enters the octahedral Al(1) sites with the dominant oxidation state 3+ (Figure 10). The presence of different Mn species with oxidation states 2+ and 3+ respectively is consistent with the average oxidation state of 2.4 measured with thermogravimetry experiments.

Correlation between the two observed Mn species and catalytic activity properties was attempted. For this purpose, turnover frequencies (TOF) referred to the bulk content of the different Mn species were derived from the results of catalytic activity tests in CH_4 combustion. TOF referred to Mn in Al(2) site was found to be almost constant on varying the overall Mn content. This suggested a possible correlation between catalyst activity and this Mn species. However, an alternative correlation was found by normalizing the catalytic activity to the surface area. Such normalized activity correlated well with the overall Mn content. No further evidence was found in favor of either these two alternatives, so that no definitive conclusion could be drawn.

Finally, the increase of Ba concentration in the mirror planes associated with incorporation of Mn^{2+} through the charge compensation mechanism described above is likely responsible for the absence of significant variation of surface area

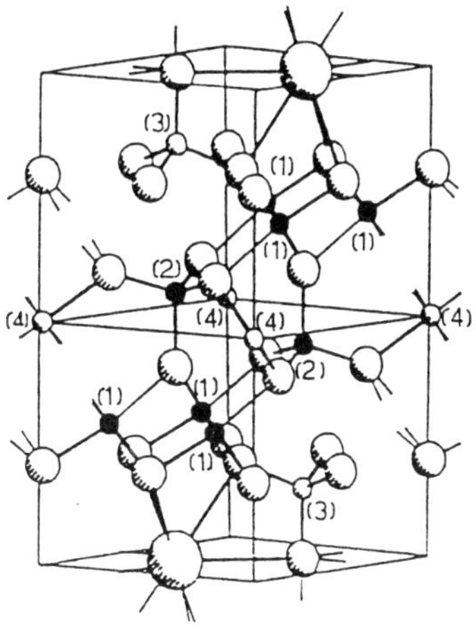

Figure 10 *Ideal semicell of Mn-substituted Ba-β-Al_2O_3. Numbers in parentheses refer to different Al sites. Bold spheres (●) represent sites of preferential Mn location.*

at low Mn loading (up to x=1). In fact, the increase of Ba concentration may balance the promotion effect of Mn on sintering up to saturation of Ba sites.

The oxidation state of Fe ions in a completely substituted $BaFe_{12}O_{19}$ catalyst with magnetoplumbite structure was studied by means of Diffuse Reflectance in the ultraviolet and visible region (DR-UV-Vis) and Mossbauer spectroscopies, and DTA-TG measurements.[24] DR-UV-Vis spectra showed the presence of both Fe^{3+} and Fe^{2+}. However no Fe^{2+} could be detected by Mossbauer and DTA-TG measurements. Accordingly the amount of Fe^{2+} is likely very low, which compares well with the average oxidation state of Fe reported in Table 4 (Fe average oxidation state = 3). DR-UV-Vis spectra also evidenced that calcination at high temperature and treatment under reaction condition result in the irreversible $Fe^{3+} \rightarrow Fe^{2+}$ reduction, which parallels a marked decrease of catalytic activity. The degree of the reduction was very low: indeed, the magnetoplumbite structure, mainly consisting of Fe^{3+} is stable up to 1300 °C. Despite its limited extent, the irreversible $Fe^{3+} \rightarrow Fe^{2+}$ reduction was hypothesized to be partly responsible for catalyst deactivation, assuming that the active iron surface species are involved in the reduction.

5 Use of Hexaaluminates in Catalytic Combustor for Gas Turbines

Different design concepts of catalytic combustors were proposed in order to meet the stringent requirements on emission control and on durability under the severe

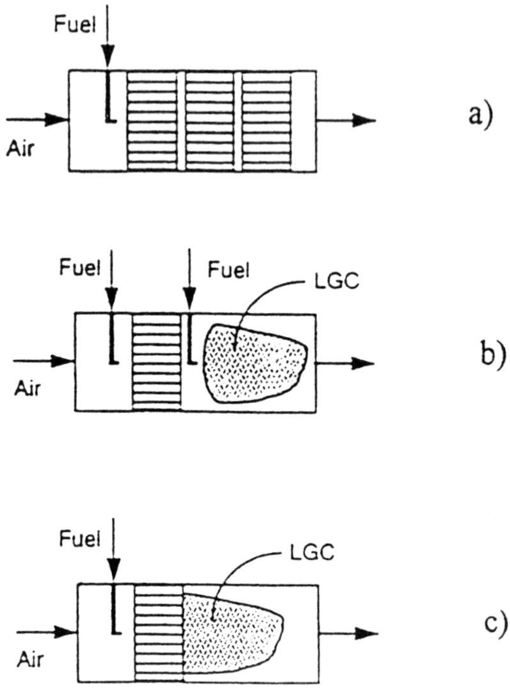

Figure 11 *Alternative configuration of catalytic combustors for gas turbines. (a) CST combustor; (b) hybrid combustor; (c) partial catalytic combustor. LGC: lean gas phase combustion*

operating conditions (high temperature and mass throughput) posed by gas turbines. Figure 11 presents the three alternative approaches that have attracted most attention.[1] In the first configuration, the so-called catalytically-stabilized thermal (CST) combustor,[34] the premixed fuel/air stream is fed to a multiple catalyst system. The combustion is first ignited at the catalyst wall and then propagates to the gas phase. In this way, the reaction proceeds until fuel consumption is complete within the catalyst section. Different types of catalyst are used along the combustor: the most active ones (typically Pd-based materials) are positioned at the entrance in order to ensure ignition at low temperature (possibly at the compressor outlet temperature); the most stable ones are used at the end region where the highest temperatures occur.

In the second configuration (hybrid combustor), only a portion of the fuel is fed to the catalyst section. The inlet air/fuel ratio is carefully controlled to limit the adiabatic reaction temperature typically below 1000 °C, and accordingly, to reduce the catalyst thermal stresses. The remaining amount of fuel is fed to a downstream homogeneous section where combustion is assisted and stabilized by the hot gas stream exiting the catalyst.

The third configuration was proposed by Catalitica Inc. and Tanaka Kikinzoku Kogyo.[48] All the fuel except that required for the pre-burner is fed to the catalyst section. The catalyst wall temperature is kept substantially below the adiabatic reaction temperature by means of a proprietary design and formulation of Pd-based catalysts. Two catalyst stages are typically present. The first one is designed to maintain a low wall temperature to make possible a stable, very high catalytic activity necessary to allow ignition at the compressor discharge temperature. The second is designed with lower catalyst activity and higher substrate temperature to provide the required gas temperature at the catalyst outlet. A subsequent homogeneous section is also present where the combustion of the fuel is completed.

The two last configurations, consisting of a low temperature catalytic combustion zone and a high temperature gas phase combustion zone, may be preferred in the future because they can allow for greater flexibility with respect to the severe operating conditions of modern high efficiency gas turbines, namely higher throughputs, combustor outlet temperatures, and pressures.

In view of the excellent thermal stability and good combustion activity observed for powder materials, hexaaluminate-type catalysts were produced in monolith form suitable for testing in prototype catalytic combustor for gas turbines fed by natural gas.[17] On the basis of optimization studies performed over powder catalysts the following composition was chosen for industrial scale-up: $Sr_{0.8}La_{0.2}Mn_1Al_{11}O_{19}$. The starting powders were manufactured using the alkoxides route. Monoliths with surface area of about 10 m^2/g and axial strengths of 20 and 30 kg/cm^2 were obtained by extrusion, upon calcination at 1200 °C and 1300 °C respectively. Since the thermal shock resistance of these materials is quite poor, the peculiar segmented catalyst configuration reported in Figure 12 was developed in order to minimize thermal stresses by loosely constraining the catalyst both in the axial and in the radial directions. This configuration consist of twelve catalyst portions designed through mathematical modeling.

Because of the activity of hexaaluminate catalysts in methane combustion is too low to ignite natural gas, a multiple catalyst system has been developed. Highly active Pd-based catalysts were used in the first segments. Intermediate segments consisted of extruded Mn-substituted hexaaluminates calcined at 1200 °C. Finally, highly stable Mn-substituted hexaaluminates calcined at 1300 °C were used as the end stages in the region where temperature reaches the maximum levels (1200 °C). This system was assembled in a 160 kW prototype combustor based on the CST configuration.[17] In full scale tests, the catalyst system attained the target performances without changes during a 215 hours continuous operation test. The catalyst system withstood crack generation in a thermal shock test in which sudden gas turbine stops were repeated five times. The following major drawbacks emerged from the prototype tests: (1) the gas throughput allowed by the tested CST configuration is too low to permit installation of the system on conventional engines configuration with very compact design;[17] (2) NO_x emissions were not as low as those levels achieved in advanced homogeneous lean premixed combustors. However, practically all NO_x

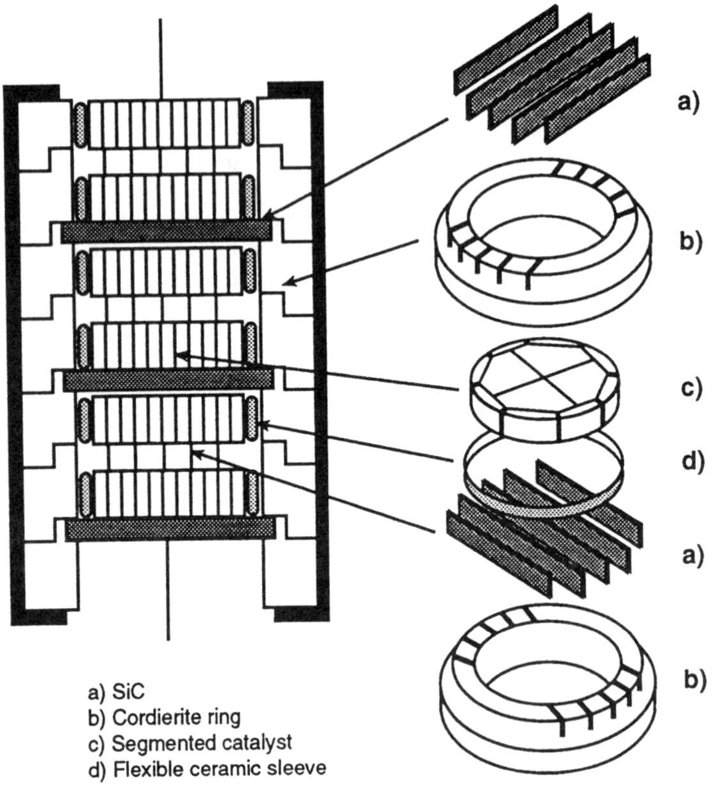

Figure 12 *Segmented catalyst configuration developed by Osaka Gas.*

production in the catalytic system originates from the diffusive preheating burner and it could be markedly reduced either when more active catalysts capable of igniting the fuel-air mixture at lower temperature or an advanced low-NO_x preheater are available.

In view of these problems, the commercialization of the system was suspended. However, these experiments definitely demonstrated durability and efficiency of hexaaluminate type catalysts under severe temperature conditions. Accordingly, their use as high-temperature end segments in combustors based on the more flexible Catalytica concept was proposed.[48] This solution could in principle provide higher gas outlet temperatures, thus reducing the residence time required to complete combustion in the homogeneous section and allowing for more compact design of the combustor.

6 Conclusions

Effective preparation methods of hexaaluminates for catalytic applications, such as the hydrolysis of alkoxides and the co-precipitation in aqueous medium, ensure high interspersion of the constituents in the precursor. This allows the formation of single phase materials with layered-alumina structure at reasonably low temperature (1100-1200 °C) and with high surface area. The hydrolysis of alkoxides was extensively studied and used for the industrial scale-up in the production of catalysts in the monolith shape. However, the co-precipitation in aqueous medium has much potential in view of the possible commercialization of these materials due to its simplicity and low cost.

The crystal structure and the sintering behavior of hexaaluminates was widely investigated. The relation of sintering resistance to anisotropic ion diffusion in the layered alumina phase was clarified to a large extent. Other evidence suggests that combustion activity is obtained through a redox mechanism associated with reversible variation of oxidation state of the transition metal ions in the structure. Mn was the best and most stable active component. However, fundamental and applied studies are needed to better clarify the redox mechanism of the reaction and how it is related to the chemical and structural features of the Mn-containing layered-alumina phase. This could also provide useful information for the development of an optimum catalyst composition,

Prototype scale experiments demonstrated durability and efficiency of hexa-aluminate type catalysts under the severe temperature conditions of industrial gas turbines fuelled by natural gas. A combination of very active Pd-based catalysts and very stable hexaaluminates and a proper catalyst aimed at minimizing the mechanical stresses due to thermal shocks was used to achieve this target.

Lab-scale tests over powder materials showed promising activity of Mn-substituted hexaaluminates in CO/H_2 combustion, suggesting their potential for use with fuels from gasification of carbon or of biomasses.

References

1. H. Arai and H. Fukuzawa, *Catalysis Today,* 1995, **26**, 217.
2. M.F.M. Zwinkels, S.G. Järås, P.G. Menon, and T. Griffin, *Catal. Rev. - Sci. Eng.*, 1993, **35**, 319.
3. E.M. Levin, C.R. Robbins and H.F. McMurdie, in "Phase Diagrams for Ceramists", ed. M.K. Reser, The American Ceramic Society, 1987, p. 97.
4. A.F. Wells, "Structural Inorganic Chemistry", Clarendon Press, Oxford, UK, 1975 p.458.
5. M. Machida, K. Eguchi and H. Arai, *J. Catal.* 1989, **120**, 377
6. R. Stevens and J.G.P. Binner, *J. Mat. Sci.*, 1984, **19**, 695.
7. M. Machida, K. Eguchi, and H. Arai, *J. Catal.*, 1987, **103**, 385.
8. M. Machida, K. Eguchi, and H. Arai, *Bull. Chem. Soc. Jpn.*, 1988, **61**, 3659.
9. H. Arai, K. Eguchi, and M. Machida, MRS Int'l. Mtg. on Adv. Mats., Material Research Society, 1989, Vol. 2, p. 243.
10. S. Chandra, *Ceram. Bull.*, 1985, **64**, 1120.
11. O.A. Kirichenko, O.V. Andrushkova, V.A. Ushakov and V.A. Poluboyarov, in

"Studies in Surface Science and Catalysis", ed. G. Poncelet, J. Martens, B. Delmon, P.A. Jacobs and P. Grange, Elsevier B.V., 1995, Vol. 91, p. 851.
12. H. Arai, Catalysts and Chemicals Inc., Far East, US Patent 4 788 174, 1988.
13. K.S. Mazdiyasni, R.T. Dolloff and J.S. Smith II, *J. Am. Chem. Soc.*, 1969, **52**, 523.
14. J.S. Smith II, R.T. Dolloff and K.S. Mazdiyasni, *J. Am. Chem. Soc.*, 1970, **53**, 91.
15. M. Machida, K. Eguchi and H. Arai, *J. Catal.*, 1990, **123**, 477.
16. K. Eguchi, H. Inoue, K. Sekizawa and H. Arai, in "Studies in Surface Science and Catalysis", ed. J.W. Hightower, W.N. Delgass, E. Iglesias and A.T. Bell, Elsevier Science B.V., 1996, Vol. 101, p. 417.
17. H. Sadamori, 3rd International Workshop on Catalytic Combustion, Amsterdam, The Netherlands, September 23-25, 1996, *Catalysis Today,* in press.
18. D.M. Lowe, M.I. Gusmann and J.G. McCarty, preprints 6th Int. Symp. on "Scientific Bases for Preparation of Heterogeneous Catalysts", September 5-8 1984, Louvain-la-Neuve Belgium, Vol. 2, p.39.
19. J.C. Debsikbar, *J. Mat. Sci.*, 1989, **24**, 3565.
20. "The Catalytic Process from Laboratory to the Industrial Plant", ed. D. Sanfilippo, 3rd Seminar of Catalysis, Italian Chemical Society, Rimini, Italy, 1994.
21. S. Matsuda, A. Kato, M. Mizumoto and H. Yamashita, Proc. 8th Int. Congress on Catalysis, Berlin 1984, Verlag Chemie Basel, Vol. IV, p. 879.
22. G. Groppi, M. Bellotto, C. Cristiani and P. Forzatti, *J. Mat. Sci.*, 1994, **29**, 3441.
23. G. Groppi, M. Bellotto, C. Cristiani, P. Forzatti and PL. Villa, *Appl. Catal. A: General*, 1993, **104**, 101.
24. G.Groppi, C. Cristiani and P. Forzatti, *J. Catal.*, 1997, **168**, 95.
25. Unpublished results from the authors' laboratories.
26. C. Cristiani, G. Groppi, P. Forzatti, E. Tronconi, G. Busca and M. Daturi, in "Studies in Surface Science and Catalysis", ed. J.W. Hightower, W.N. Delgass, E. Iglesias and A.T. Bell, Elsevier Science B.V., 1996, Vol. 101, p. 473.
27. B.W.L. Jang, R.M. Nelson, J.J. Spivey, M. Ocal, R. Oukaci and G. Marcelin, 3rd International Workshop on Catalytic Combustion, Amsterdam, The Netherlands, September 23-25, 1996, *Catalysis Today,* in press.
28. M. Berg, S. Järås, *Catalysis Today*, 1995, **26**, 223.
29. G. Busca, C. Cristiani, P. Forzatti and G. Groppi, *Catalysis Letters*, 1995, **31**, 65.
30. P. Burtin, J.P. Brunelle, M. Pijolat, and M. Sustelle, *Appl. Catal.*, 1987, **34**, 225.
31. M.L.F. Johnson, *J. Catal.*, 1990, **123**, 245.
32. J. G. Mc Carty, 3rd International Workshop on Catalytic Combustion, Amsterdam, The Netherlands, September 23-25, 1996, *Catalysis Today,* in press.
33. H. Arai and M. Machida, *Catalysis Today,* 1991, **10**, 81.
34. L.D. Pfefferle and W. C. Pfefferle, *Catal. Rev. Sci. Eng.*, 1987, **29**, 219.
35. A. Nakhjavan, P. Bjorbom, M. Zwinkels and S.G. Jaras, *Chem. Eng. Sci.*, 1995, **50**, 2255.
36. G. Groppi, E. Tronconi and P. Forzatti, *Appl. Catal A: General*, 1996, **138**, 177.
37. M. Machida, K. Eguchi and H. Arai, *J. Am. Ceram. Soc.*, 1988, **71**, 1142
38. M. Machida, T. Shiomitsu, K. Eguchi, H. Arai and Y. Shimizu, *J. Solid State Chem.*, 1991, **95**, 220.
39. M. Machida, A. Sato, T Kijima, H. Inue, K. Eguchi and H. Arai, *Catalysis Today*, 1995, **26**, 239.
40. G. Groppi, F. Assandri, M. Bellotto, C. Cristiani and P. Forzatti, *J. Solid State Chem.*, 1995, **114**, 326.
41. S. Kimura, E. Bannai and I. Shindo, *Mat. Res. Bull.*, 1982, **17**, 209.
42. N.Ivy, Z.Inoue, S.Takekawa and S. Kimura, *J. Solid State Chem.*, 1984, **52**, 66.
43. N.Ivy, Z.Inoue, S.Takekawa and S. Kimura, *J. Solid State Chem*, 1985, **60**, 41.
44. M. Bellotto, G. Artioli, C. Cristiani, P. Forzatti and G. Groppi, 3rd International Workshop on Catalytic Combustion, Amsterdam, The Netherlands, September 23-25, 1996, *Catalysis Today,* in press.
45. P. Batti and F. Di Tella, *Ann. Gazzetta Chimica Italiana*, 1967, **57**, 74.
46. G. Groppi, A. Belloli, E. Tronconi and P. Forzatti, *Catalysis Today*, 1996, **29**, 403.

47. F. Laville, D. Gourier, A.M. Lejus and D. Vivien, *J. Solid State Chem.*, 1983, **49**, 180.
48. R.A. Dalla Betta, N. Ezawa, K. Tsurumi, J.C. Schlatter S.G. Nickolas, US Patent 5183401, 1993.

5

Catalytic Conversions in Water. An Environmentally Benign Concept for Heterogenization of Homogeneous Catalysis*

BY GEORGIOS PAPADOGIANAKIS† AND ROGER A. SHELDON

1 Introduction

Catalysis is the key to efficiency of chemical reactions and plays a major role in the development of alternative environmentally benign manufacturing technologies that are more energy efficient, produce a minimum amount waste and avoid the use of toxic and/or hazardous reagents and solvents.[1] In 1993, it was estimated that the manufacture of 80% of chemical products involved catalysis[2] and according to the *Pimentel report*[3] an estimated 20% of the GNP (Gross National Product) of the USA is generated through the use of catalytic processes.

Catalysts, including enzymes, are conveniently divided into two categories: heterogeneous and homogeneous. In heterogeneous systems the catalyst is usually a solid and the reactants and products are contained in a liquid or gas phase.[4] In homogeneously catalysed reactions the catalyst and the reactants form one phase, and no phase boundary exists. The major advantages and disadvantages of homogeneous and heterogeneous catalysis are summarized in Table 1.[5-8]

Table 1 *Comparison between homogeneous and heterogeneous catalysis*

	Homogeneous	*Heterogeneous*
Selectivity	very high	often a problem
Activity	all metal atoms are accessible	only surface atoms
Reaction conditions	mild (50-200°C)	forcing (often > 250°C)
Diffusion problems	none	sometimes
Mechanistic understanding	NMR, IR etc can be applied under reaction conditions	more difficult
Reproducibility	no problem	often difficult
Catalyst separation	cumbersome	easy

* Part 7. For Part 6 of this series see Ref. 452.

† Present address: University of Athens, Laboratory of Industrial Chemistry, Panepistimiopolis - Zografou, 15771 Athens, Greece.

Characteristic features attributed to homogeneous transition metal catalysts are their superior selectivity and activity compared with their heterogeneous counterparts. However, a serious shortcoming of homogeneous catalysis is the cumbersome separation of the catalyst from reaction products and the quantitative recovery of the catalyst in an active form.

There have been many attempts to heterogenize homogeneous catalysts by attachment to organic or inorganic supports. These approaches have not resulted, to our knowledge, in any commercially viable process for a variety of reasons, such as poor catalyst productivities, leaching of the metal, irreproducible activities and selectivities and degradation of the polymer support. At least one industrial process (pilot plant stage) uses membranes for the separation of products from the homogeneous catalyst namely the rhodium catalysed hydroformylation of dicyclopentadiene to tricyclodecane dialdehyde in organic solvents.[287,307]

Another approach is to separate the products from the homogeneous catalyst using a two phase liquid system. For example, this method is used in the oligomerization step of the Shell Higher Olefins Process for the manufacture of linear α-olefins.[5,9-11,330] A polar nickel catalyst containing a P-O chelate ligand is dissolved in a polar solvent e.g. 1,4-butanediol, which is immiscible with higher α-olefins, and recovery of the catalyst is easily achieved by simple phase separation.

An environmentally attractive approach is to use water, which is highly polar and therefore immiscible with most organic products, as the second phase. The water soluble catalyst resides and operates in the aqueous phase and, after the reaction, quantitative catalyst recovery is easily achieved by simple phase separation from organic products. The concept of two immiscible liquid phases was first suggested by Manassen[27,28] in 1972 and received more attention after the interesting review of Bailar[8] entitled "Heterogenizing Homogeneous Catalysts" in 1974.[29] In the same year Kuntz[30] of Rhône-Poulenc prepared the sodium salt of trisulfonated triphenylphosphine (tppts, see Table 2, **1**, n=0) and hydroformylated propene using the $RhH(CO)(tppts)_3$ catalyst in an aqueous/organic two phase system. Subsequently Cornils and coworkers[31] of Ruhrchemie AG (today Hoechst AG Werk Ruhrchemie) developed, after further optimization of the synthesis and purification of tppts,[32,33] an industrial process for the hydroformylation of propene[34] in a two phase system. The Ruhrchemie/Rhône-Poulenc (RCH/RP) process is a benchmark in the field of organometallic catalysis in aqueous media and its success stimulated the application of this concept to a broad spectrum of catalytic reactions. Aqueous/organic two phase systems have the following advantages: (i) numerous steps in classical processes are rendered superfluous and process engineering is enormously simplified, resulting in substantial energy savings and conservation of resources,[35-38,55] (ii) lower emissions and minimum environmental pollution,[35-40] and (iii) they obviate the need for organic solvents and water is a safe, non-toxic, inexpensive, widely available and environmentally attractive solvent.[35]

More recent conceptual advances in the liquid/liquid biphasic systems include the novel fluorous biphase system (FBS)[12-14,70-73] and the use of ionic liquids[15-22] as solvents. Catalysis in supercritical fluids[23-26] is another environmetally attractive approach. The FBS concept is based on the limited miscibility of common organic solvents and products with the fluorous phase which consists of an FBS compatible catalyst dissolved in a fluorinated solvent such as a perfluorinated alkane, ether or amine. For example, in hydroformylation reactions of α-olefins in toluene using $RhH(CO)\{P[(CH_2)_2(CF_2)_5CF_3]_3\}_3$ catalysts in $C_6F_{11}CF_3$ as the fluorous phase no carry-over of rhodium to the hydrocarbon phase took place.[12-14,70-73] Room temperature or low-melting ionic liquids, e.g. 1,3-dialkylimidazolium,[15-21] 1-alkylpyridinium,[20] or tetraalkylammonium[22] salts of BF_4^-, PF_6^-, $AlCl_4^-$ etc. anions have been used as solvents to perform catalytic reactions in two-phase systems. They are immiscible with most organic solvents and can dissolve catalysts either because of their ionic character or by using polar ligands.

Various reviews[31,35-38,41-69,221,224] have been devoted to aqueous organometallic chemistry and catalysis in general and to specific topics such as biomedical applications of some water soluble phosphines[41] or reactions catalysed by water soluble palladium complexes.[43,44,55,56] A related review[69] describes the synthesis of organometallic clusters in aqueous solutions. The state of the art and the potential of organometallic chemistry and catalysis in aqueous media were critically evaluated at a NATO Advanced Research Workshop in 1994[70-72] and at the 209th Meeting of the American Chemical Society (Anaheim, CA) in 1995.[221,223] Furthermore, a special issue of the *Journal of Molecular Catalysis A: Chemical* is entirely devoted to *Catalysis in Water*[225] and a chapter of a forthcoming volume of *Inorganic Syntheses* describes water soluble ligands and their complexes.[226]

This review primarily focuses on the numerous reactions catalysed by water soluble transition metal complexes with emphasis on their implications for development of new environmentally benign processes in aqueous media based on the easy and quantitative separation of products from the catalyst as well as the avoidance of organic solvents. Moreover it contains, to our knowledge, the first comprehensive overview of water soluble ligands which play a key role for the development of efficient organometallic catalysis in aqueous media.

2 Water Soluble Ligands

Aqueous phase organometallic catalysis requires a modification of classical organometallic complexes, usually by ligand variation, in order to render them water soluble. Water solubility of appropriate ligands is achieved by introducing charged or polar substituents. On the other hand numerous examples of aqueous phase organometallic catalysis are known where the water solubility of transition metal complexes does not derive from the water solubility of elaborate ligands but from specific water-metal interactions e.g. with $(Me_3P)_3Ir(H)(H)Cl$ complexes,[602,603] from aqua ligands e.g. in $[Ru(H_2O)_6]^{2+}$ catalyst precursors[604,605] or from CN^- and CO ligands e.g. in $[Co(CN)_5]^{3-}$ or $[Rh(CO)_4]^-$ catalytic systems.[30,549,550a,595]

In this overview the water soluble ligands are divided into sulfonated phosphines (Table 2), carboxylated phosphines (Table 3), phosphines containing hydroxy and ether functionalities (Table 4), phosphines with amino, ammonium, phosphonium, phosphonate and phosphate moieties (Table 5) and miscellaneous systems (Table 6) such as sulfonated amines, phthalocyanines, porphyrins, triphenylarsine, phosphole, and thioethers or sulfone-based phosphines, nitrogen containing ligands mainly used as contrast agents for Magnetic Resonance Imaging, and some tenside ligands.

The water solubility of tppms (Table 2, **3**, n=2) is 80 g/l at 20°C[30] (according to another source 200 g/l at room temperature,[215] the lithium salt of tppms exhibits a surprisingly high water solubility of 1350 g/l[334]) and of tppts 1100 g/l at 20°C.[30] The trisulfoalkylated tris(2-pyridyl) phosphine **15** (Table 2; n=3) is even more water soluble (1350 g/l at 27°C).[115] However, to our knowledge, the most water soluble (2500 g/l) functionalised phosphine ligand is the phosphine **128** n=0; R= CH_2SO_3Na (Table 6) containing sulfonate and thioether groups.[333] Surprisingly, the phosphine **85** R=H; X=Br (Table 5) is also more water soluble than tppts, namely 1450 g/l at 20°C.[227] The phosphine **85** is obtained via protonation of the $-NMe_2$ group of the basic phosphine **84** (n=2, R= Me) using $PPh_3{\cdot}HBr$.[227,289]

Because tppts is so far the best water soluble ligand, in terms of favorable price/performance ratio, availability and stability we briefly describe the synthesis of this ligand. For the synthesis of the other water soluble ligands the reader is referred to the literature cited in each Table.

The first water soluble sulfonated triphenylphosphine ligand, tppms, was prepared in 1958, by Ahrland and Chatt,[98-100] by direct sulfonation of triphenylphosphine with 20%-65% oleum (SO_3/H_2SO_4). Kuntz[30] sulfonated triphenylphosphine using 20% oleum and longer reaction times than for the preparation of tppms. After hydrolysis and addition of NaOH, tppts was the main product (approximately 80%) together with the disulfonated and monosulfonated phosphines, the corresponding phosphine oxides[77] and sulfides as impurities. Ruhrchemie AG[32-33] developed a method for preparation of tppts, with a purity of >95% on a ton scale, by sulfonation of triphenylphosphine with 30% oleum at 20°C for 24 hours. After careful hydrolysis of the reaction mixture, the thus-formed sulfonic acids of triphenylphosphine are extracted with a solution of a water insoluble amine, usually tris(isooctyl)amine, dissolved in toluene. Tppts is obtained after pH controlled selective re-extraction of the organic phase containing the quaternary ammonium salts of sulfonated triphenylphosphine with a 5% aqueous NaOH solution. The main impurity is tppds (Table 2, **2**, n=1). By optimizing the reaction conditions and pH range for the re-extraction of the organic phase with aqueous NaOH it was possible to obtain tppts, tppds or tppms with purity higher than 99%.[305,306] Recently, a new method of sulfonation of phosphines claims that addition of orthoboric acid to the sulfonation mixture suppresses or eliminates phosphine oxide formation.[82-85] Thus, using the $H_2SO_4/B(OH)_3/SO_3$ sulfonation mixture tppds was selectively obtained from PPh_3 and the highly sulfonated phosphines **5** (R= CH_3, OCH_3; n=0) and **25** (Table 2) and the first sulfonated phosphole **116** (Table 6) were prepared for the first time without contamination by oxidation products.[82-85]

Table 2 *Water soluble sulfonated phosphines* ($AR = C_6H_4$-*m*-SO_3Na)[a,b]

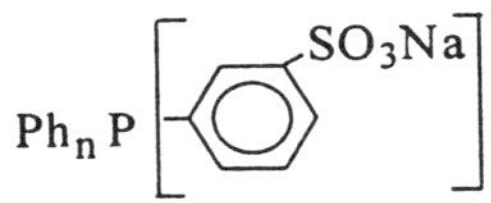

1 (tppts) n = 0
2 (tppds) n = 1
3 (tppms) n = 2

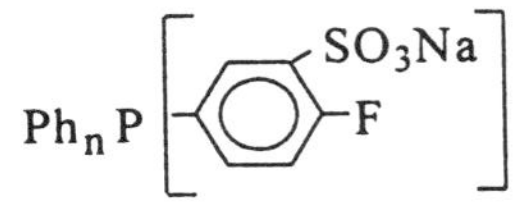

n = 0, 1, 2

4

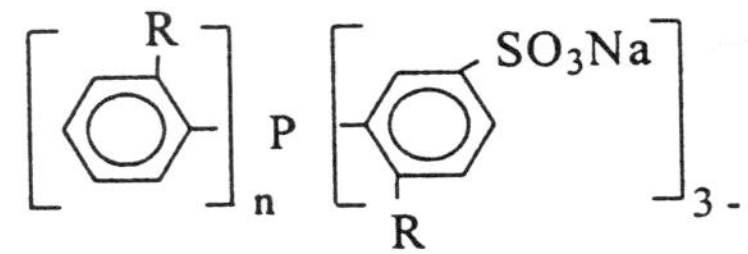

R = Me, Et ; n = 0, 1, 2
R = OCH_3 ; n = 0

5

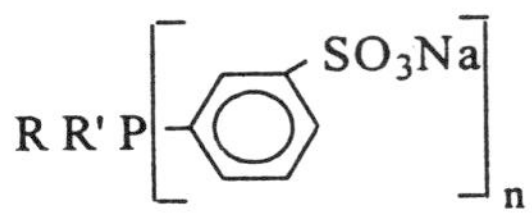

n=0, R = Me, R' = Ar, Et, Ph
n=2, R = R' = n-Hexyl

6

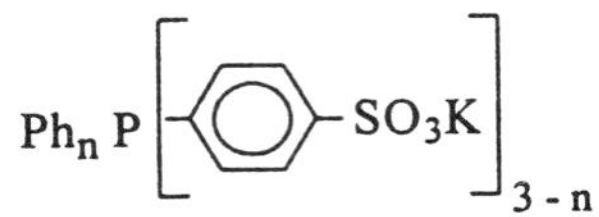

n = 0, 1, 2

7

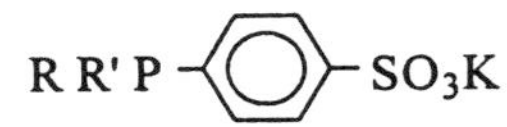

R = Ph, R' = Me ; R = R' = 3-Py
R = 2,4,6- $Me_3C_6H_2$, R' = CH_2Ph
R = 2,4,6- i $Pr_3C_6H_2$, R' = Me, Ph

8

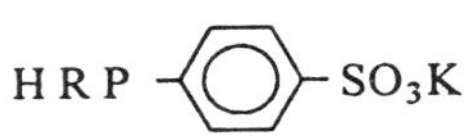

R = 2,4,6-$Me_3C_6H_2$;
2,4,6- i $Pr_3C_6H_2$

9

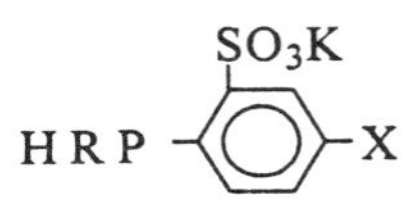

R = Ph ; X = H
R = Ph ; X = SO_3K
R= 2,4,6-$Me_3C_6H_2$; X = SO_3K

10

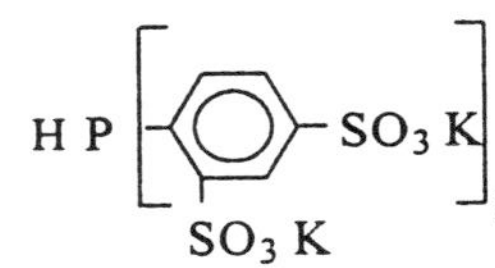

11

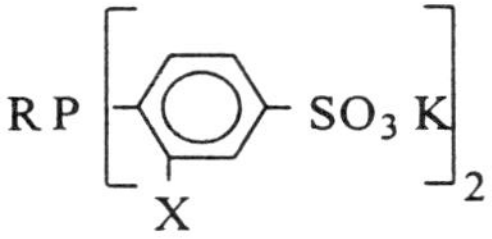

R = Ph, nBu, CH_2Ph; X = SO_3K
R = 2,4,6- $Me_3C_6H_2$; X = H

12

R = R' = Ph ; R = Ph, R' = Me
R = Ph , R' = CH_2Ph ; R = Ph ; R' = $C_{12}H_{25}$

13

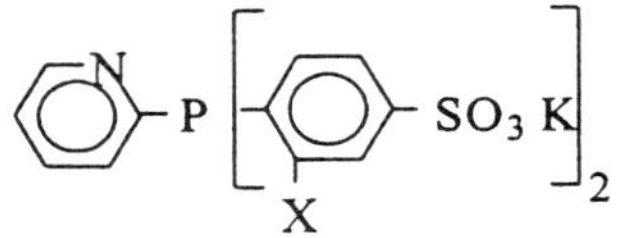

X = H , SO_3K

14

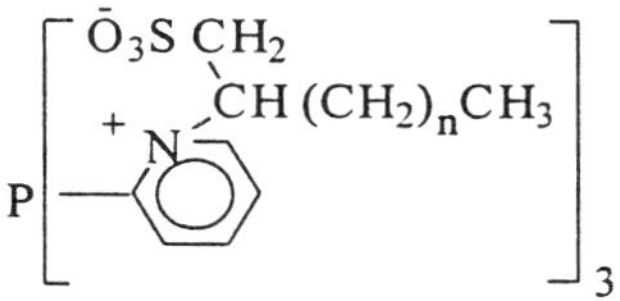

n = 0, 3, 5, 7, 9, 11

15

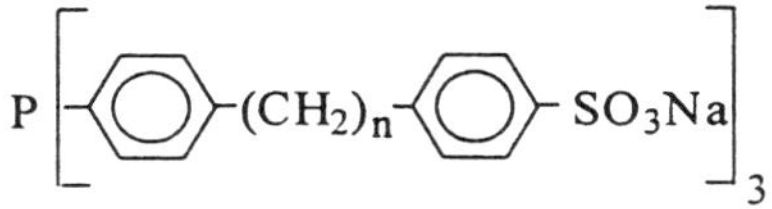

n = 3, 6

16

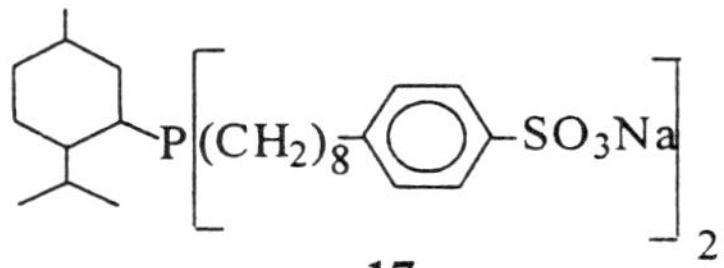

17

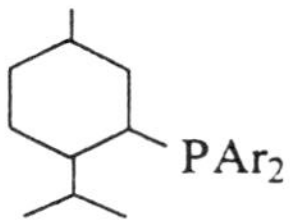

18

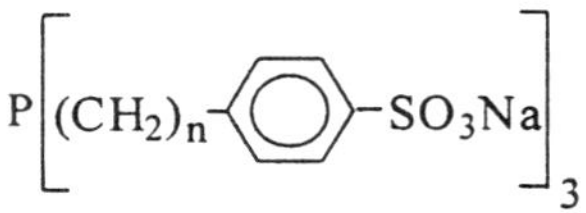

n = 1, 2, 3, 6

19

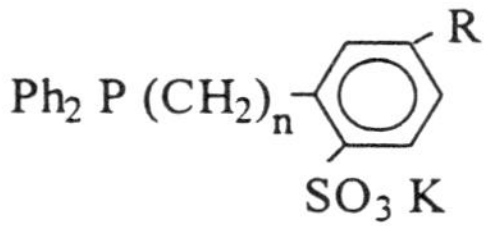

n = 0, 1 ; R = H
n = 0 ; R = Me

20

$R_2P(CH_2)_nSO_3Na$

R = Ph ; n = 2, 3 ,4
R = Menthyl ; n = 2 ,3 ,4
R = Cyclohexyl; n = 2

21

Table 2 *(contd)*

$PhP[(CH_2)_4SO_3Na]_2$

22

$Ph_2P(CH_2)_2C(=O)NHC(CH_3)_2CH_2SO_3Li$

23

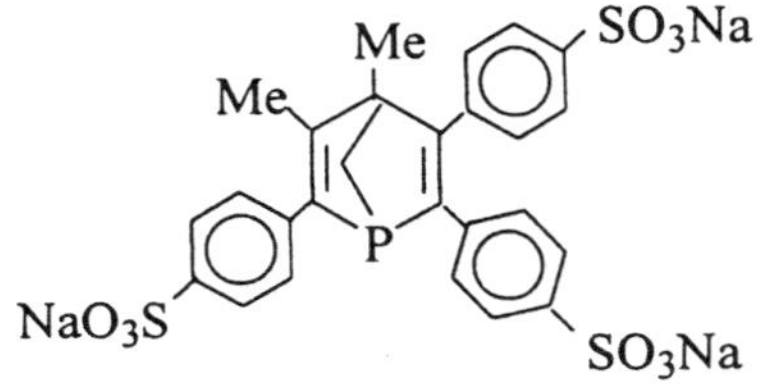

24

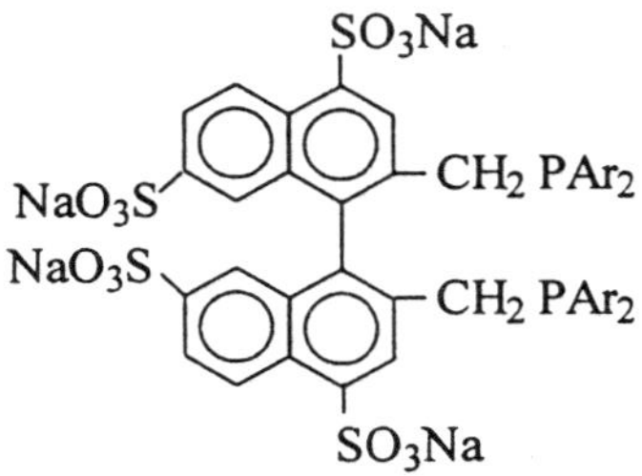

25

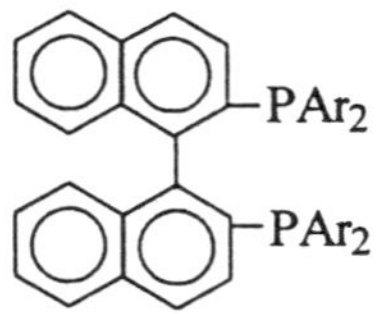

26

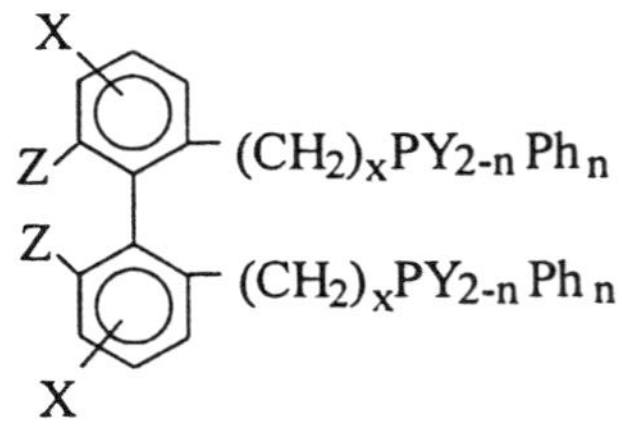

X= SO_3Na, Z= H, Y=Ar, x = 1, n = 0,1

X= H, Z = OMe, Y= C_6H_4-*p*-SO_3Na, x = 0, n= 0

27

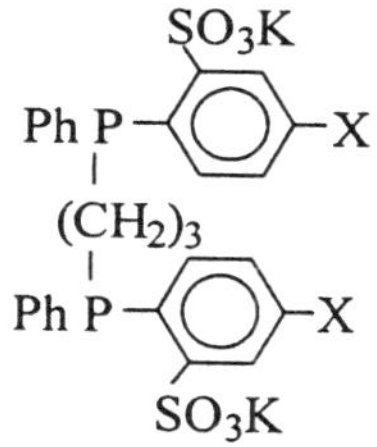

$X = SO_3K, H$

28

$Ph_m Ar_{2-m} P (CH_2)_x P Ph_n Ar_{2-n}$

$x = 2, 3, 4$

$m = 0,\ n = 0\ ;\ m = 0,\ n = 1$
$m = 1,\ n = 1\ ;\ m = 1,\ n = 2$

29

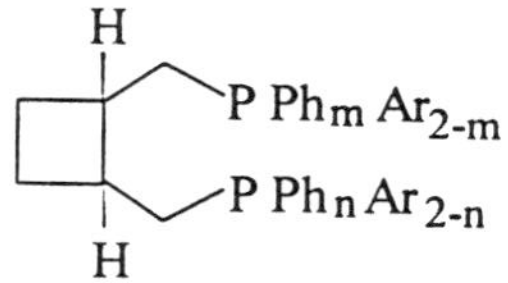

$m = 0,\ n = 0\ ;\ m = 0,\ n = 1$
$m = 1,\ n = 1\ ;\ m = 1,\ n = 2$
$m = 1,\ n = 1,\ Ar = nBuSO_3Li$

30

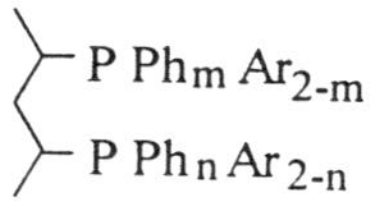

$m = 0,\ n = 0\ ;\ m = 0,\ n = 1$
$m = 1,\ n = 1\ ;\ m = 1,\ n = 2$

31

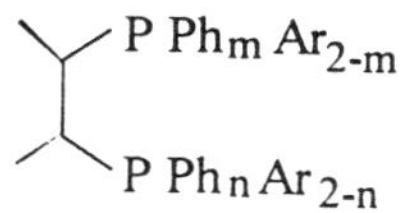

$m = 0,\ n = 0\ ;\ m = 0,\ n = 1$
$m = 1,\ n = 1\ ;\ m = 1,\ n = 2$

32

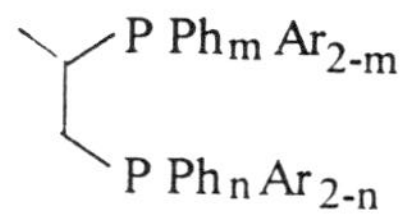

$m = 0,\ n = 0\ ;\ m = 0,\ n = 1$
$m = 1,\ n = 1\ ;\ m = 1,\ n = 2$

33

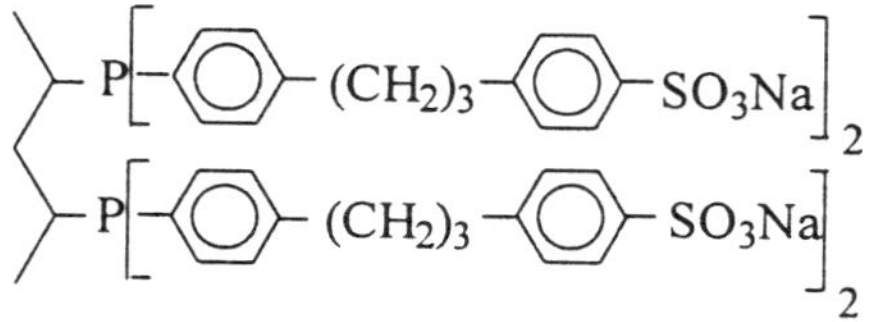

34

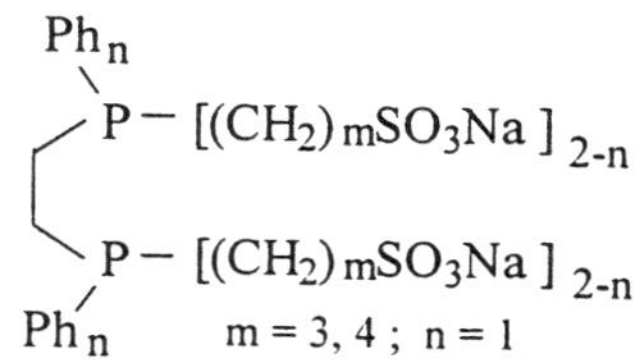

$m = 3, 4\ ;\ n = 1$
$m = 3\ ;\ n = 0$

35

Table 2 *(contd)*

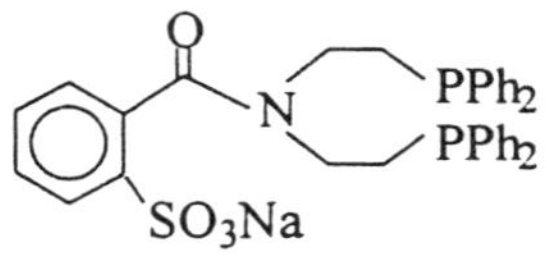

36

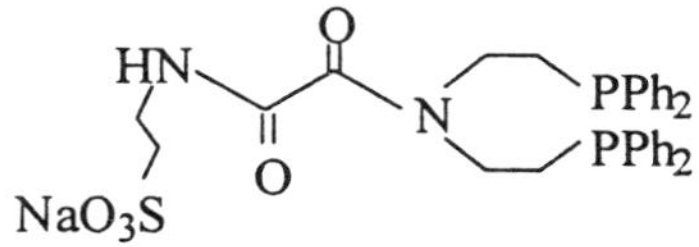

37

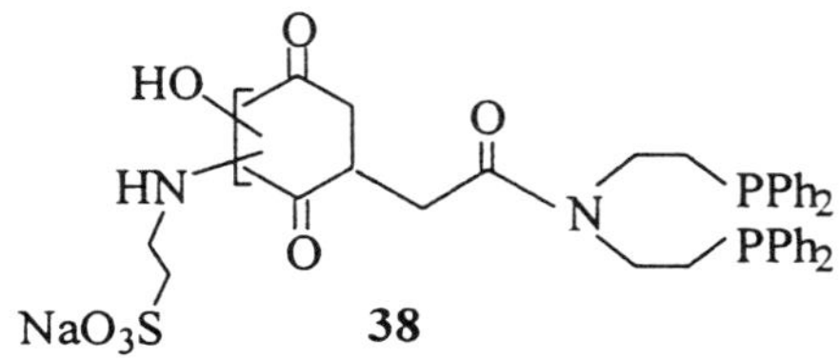

38

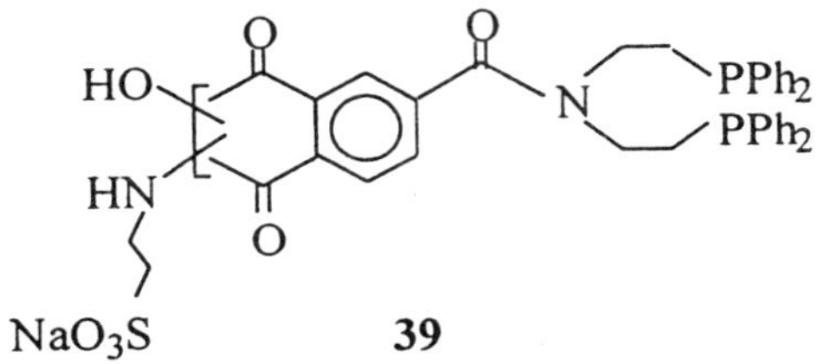

39

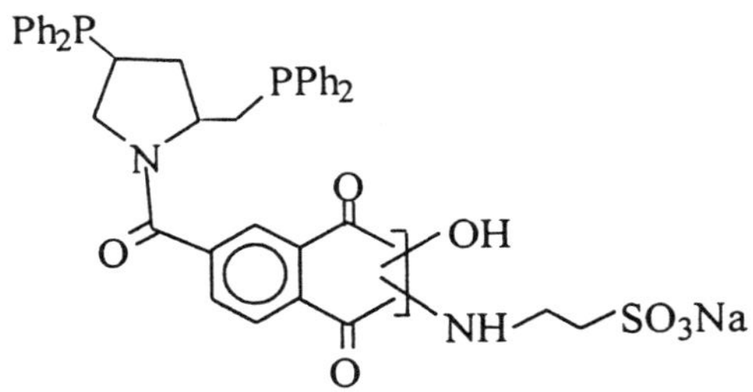

40

[a] **References:** **1**[1, 13, 15-17, 28, 30, 31-38, 41-43, 45-58, 60-67, 70-72, 74-97, 109-111, 114-122, 132, 141, 155, 165, 172, 173, 187-189, 221, 224, 233, 246, 250, 252, 259, 261, 305, 320, 414, 417, 505, 550b]; **2**[31-37, 41, 48, 49, 53, 77, 82-85, 172, 173, 305, 516]; **3**[15-17, 29-37, 41, 44, 46, 48, 49, 51, 53, 55, 56, 59-63, 65, 66, 68, 70, 72, 74, 77, 80, 87, 94, 95, 98-106, 111, 114, 117, 118, 123, 141, 150, 166, 172, 173, 187, 188, 191, 203, 214, 215, 221, 224, 246, 262, 265, 266, 320, 321, 330, 334, 413, 509]; **4**[75, 76, 107, 305]; **5**[48, 82, 84, 85, 108, 221]; **6**[109, 114, 172, 563]; **7**[53, 61, 66, 110-113, 164, 221]; **8**[110-112, 114]; **9**[114]; **10**[114]; **11**[111, 114]; **12**[111, 114]; **13**[111, 114]; **14**[110-112]; **15**[31, 28, 35, 37, 42, 48, 52, 53, 61, 62, 115-119, 165, 200, 230, 238, 320]; **16**[31, 48, 81, 117, 118, 149, 320]; **17**[118, 119, 221]; **18**[61, 66, 144, 172]; **19**[31, 48, 53, 118, 119, 120-122]; **20**[53, 123, 124, 125, 333, 335, 524]; **21**[51, 53, 61, 118, 123-129, 309, 331, 335]; **22**[127]; **23**[130, 131, 221]; **24**[31, 35, 37, 38, 48, 49, 61, 132, 133, 224, 320, 374]; **25**[31, 35, 37, 47-49, 82, 85, 132, 133, 221, 224, 320, 550b]; **26**[1, 35, 50, 135-138, 144]; **27**[31, 35, 37, 38, 48, 49, 51, 52, 53, 61, 81, 117, 132, 134, 139, 221, 222, 224, 320, 375]; **28**[114]; **29**[53, 61, 114, 139-141, 172, 173, 187, 221, 246, 263, 451]; **30**[51, 53, 61-63, 66, 87, 111, 114, 139-141, 143-147, 172, 173, 308]; **31**[51, 53, 61-63, 66, 87, 114, 139, 140, 142-148, 172, 173, 587]; **32**[51, 53, 61-63, 66, 111, 114, 123, 139, 140, 143-147, 172, 173]; **33**[53, 61, 62, 87, 139, 143, 144]; **34**[149, 221]; **35**[53, 123, 126, 228]; **36**[53, 62, 66, 139, 150, 151]; **37**[62, 66, 150]; **38**[51, 62, 66, 150, 151]; **39**[53, 61, 62, 66, 150-152]; **40**[53, 62, 63, 66, 153]

[b] Various other counter cations such as Li^+, Cs^+, Ca^{2+}, Ba^{2+}, Al^{3+}, NH_4^+, NHR_3^+, NR_4^+ (R= C_1- C_{24}) etc. have been described for sulfonated phosphines.

Table 3 *Water soluble carboxylated phosphines*[a]

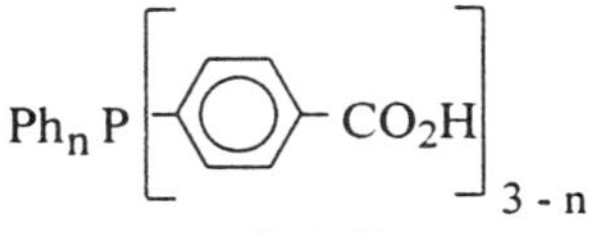

n = 0, 1, 2

41

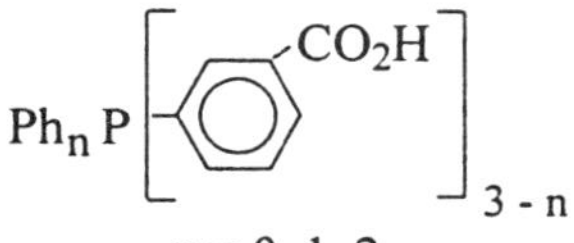

n = 0, 1, 2

42

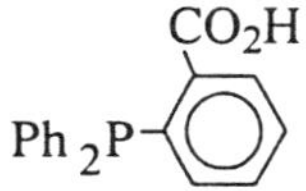

43

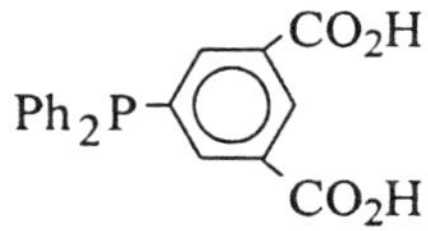

44

$Ph_2P(CH_2)_nCOOH$

n = 1, 2, 3, 4, 5, 7, 9, 11

45

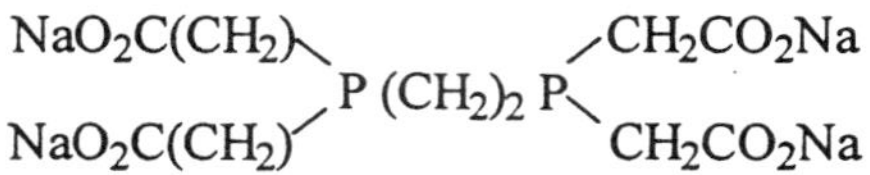

46

$R_nP[(CH_2)_xCOOR']_{3-n}$

R= Me, Et, Ph ; R' =H, Na, Me
n= 0, 1 ; x = 1, 2

47

$R_nP(CH_2OCOMe)_{3-n}$

R= Me, Et, Ph ; n= 0, 1, 2

47a

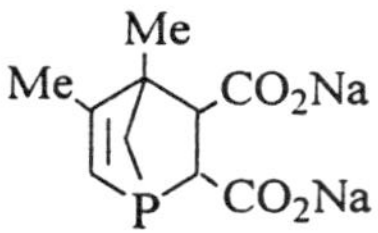

48

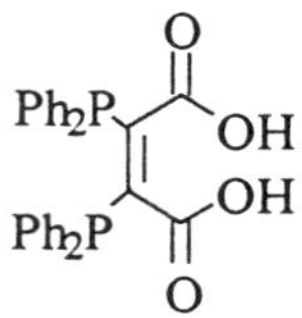

49

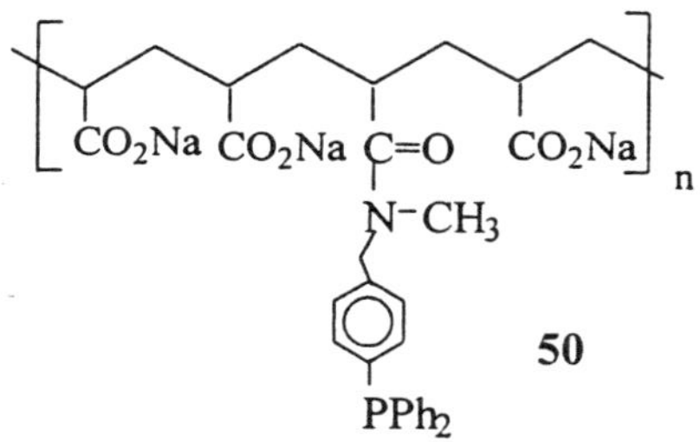

50

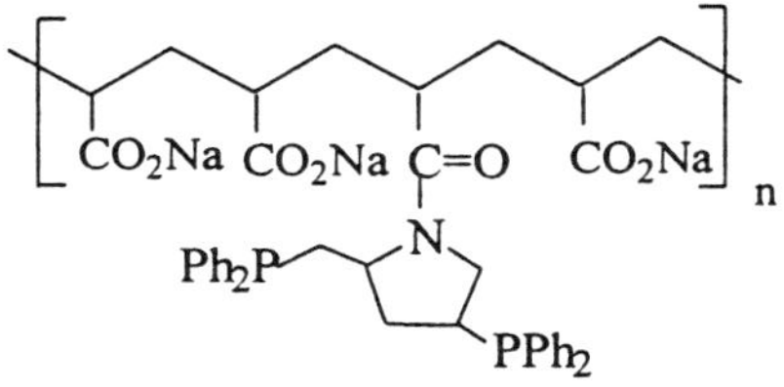

51

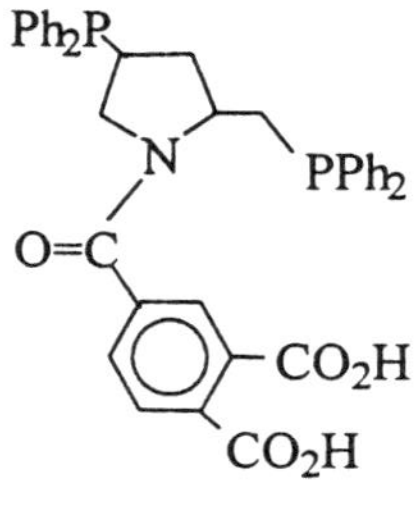

52

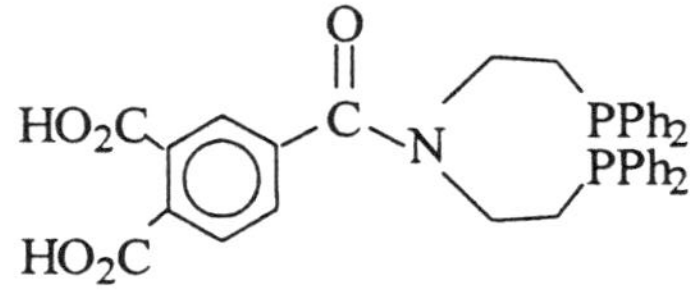

53

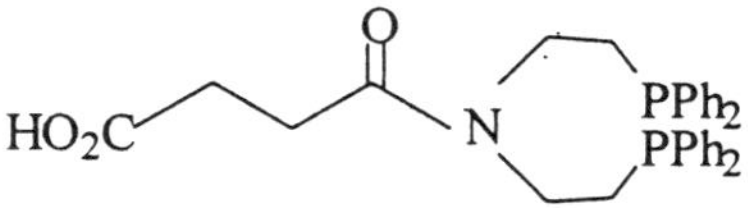

54

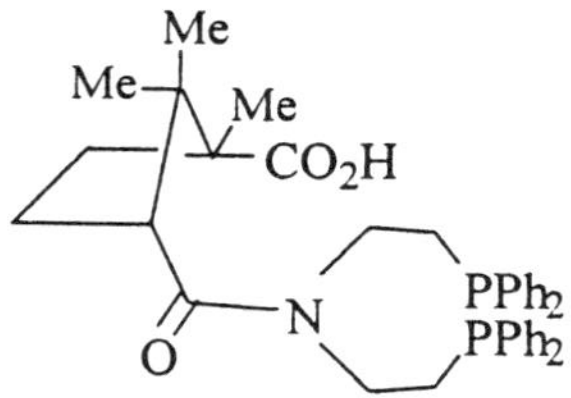

55

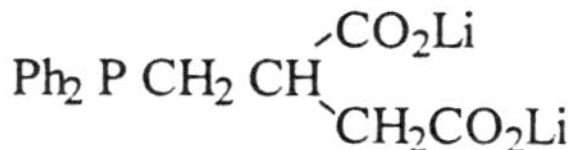

56

---

[a] **References: 41**[53, 61, 87, 154, 155, 156-160, 164, 165, 229, 238, 320]; **42**[31, 37, 48, 53, 61, 154, 160, 164, 215, 229, 259, 330, 525]; **43**[61, 161, 163, 330]; **44**[164]; **45**[5, 9-11, 31, 37, 48, 53, 61, 87, 123, 162, 166-169, 181, 183, 185, 210, 211, 229, 330]; **46**[61, 170, 171, 174]; **47**[31, 37, 48, 53, 62, 171, 175-182, 184, 194]; **48**[61, 186]; **49**[51, 61, 139, 187, 246]; **50**[188, 501]; **51**[189, 501]; **52**[51, 53, 62, 63, 66, 153]; **53**[53, 62, 66, 123, 150, 151]; **54**[62, 66, 123, 151, 190]; **55**[53, 62, 66, 68, 151, 190]; **56**[130, 221]

Table 4 *Water soluble phosphines containing hydroxy and ether functionalities*[a]

$P[(CH_2)_nOH]_3$

n = 1, 3

57

$R_2P(CH_2)_nOH$

n = 2, 3 ; R = Et, Hexyl, Ph

58

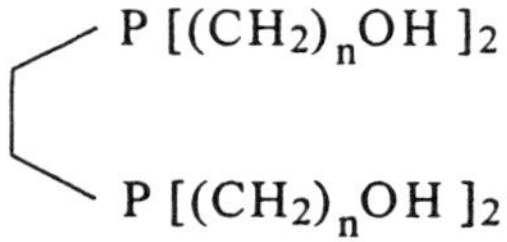

n = 1, 3

59

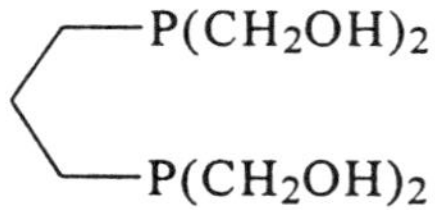

60

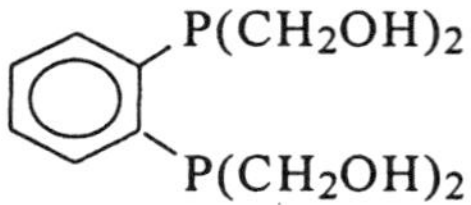

61

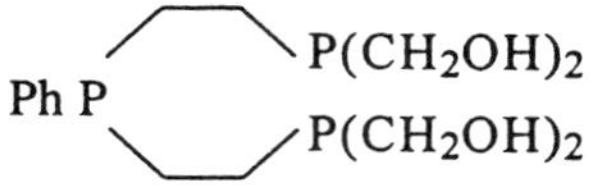

62

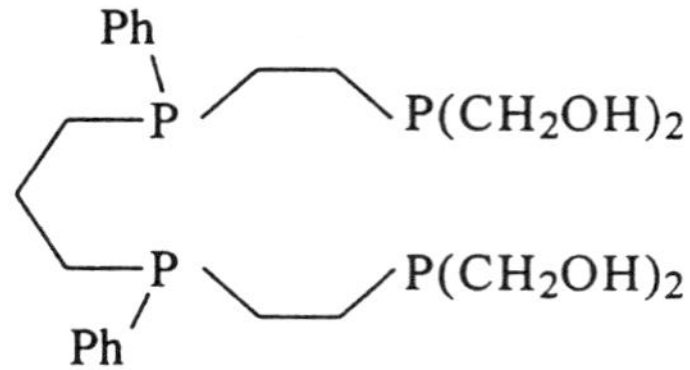

63

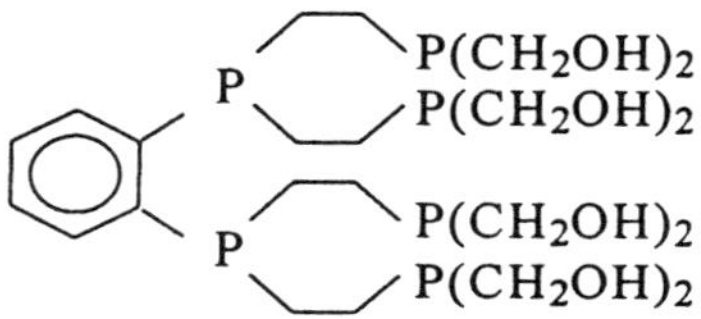

64

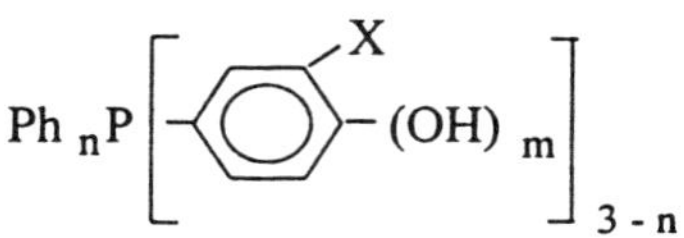

n=0, 2; m=0, 1; X=H, OH, COOH

65

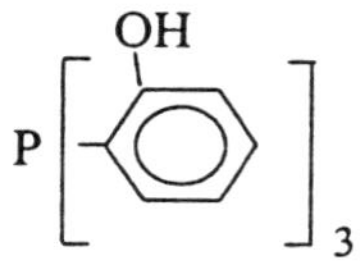

66

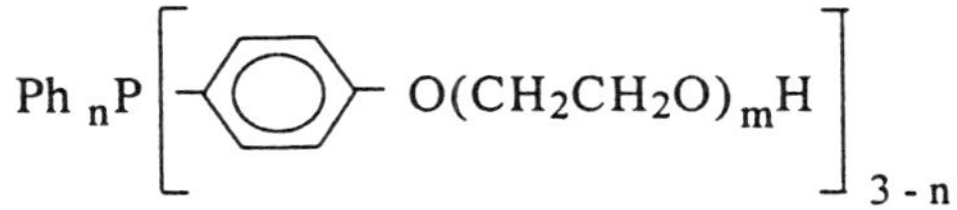

n = 0, 1, 2;   m = 3 - 25

67

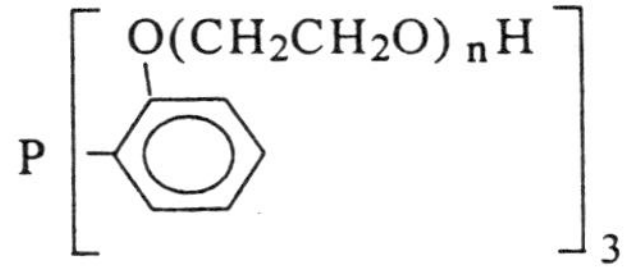

n = 3 - 14

68

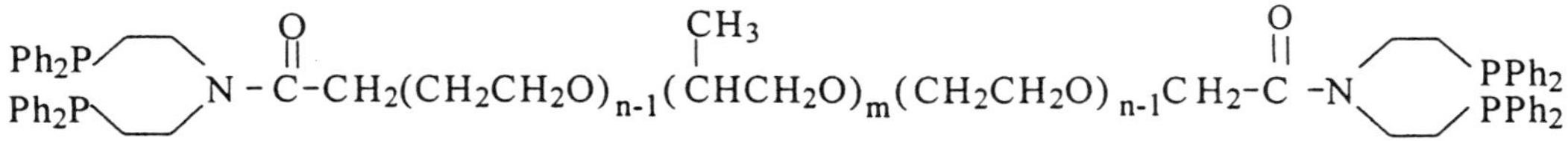

n = 11 ;   m = 34

69

$Ph_2P(CH_2CH_2O)_{n-1}CH_2CH_2PPh_2$

n = ca. 2000

70

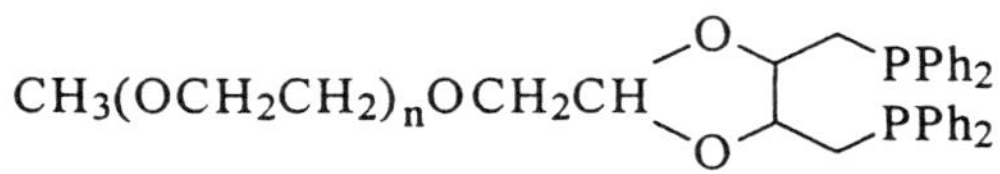

n = (5), 16, 42

71

Table 4 *(contd)*

$CH_3(OCH_2CH_2)_nOCH_2CH(PPh_2)CH_2PPh_2$

n = 5, 18

72

$Ph_nP[(CH_2)_3O(CH_2CH_2O)_mR]_{3-n}$

n=0, 1, 2; m=1, 2, 3; R=H, Me

73

$Ph_nP[(CH_2)_3OCH_2CH(OH)CH_2OH]_{3-n}$

n=1, 2

74

$Ph_nP[(CH_2)_3O\ sugar]_{3-n}$

n=1, 2

Sugar= glucose, fructose etc.

75

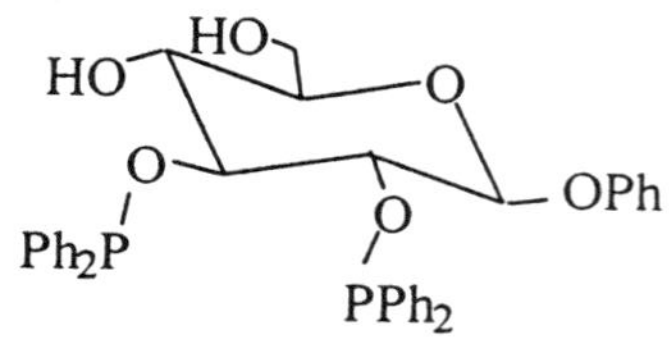

75a

$Ph_nP[(CH_2)_x(CH_2CH_2O)_mR]_{3-n}$

n=0,1; x=0,1; m=1,2,3; R=Me, Bu

76

$P[CH(CH_3)(CH_2CH_2O)_nCH_3]_3$

n=1, 2, 3

77

$PhP[CH_2(CH_2CH_2O)_nCH_3]_2$

n=1, 2

78

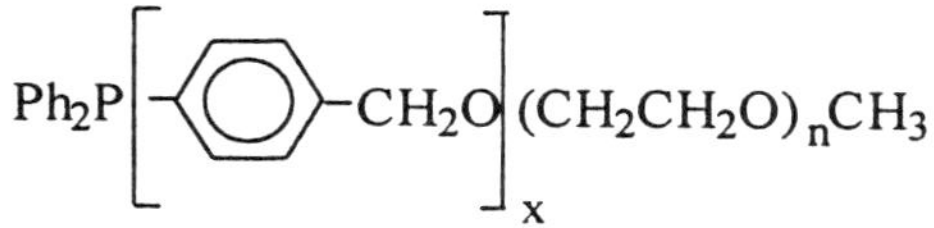

x= 0, 1; n=12, 16

79

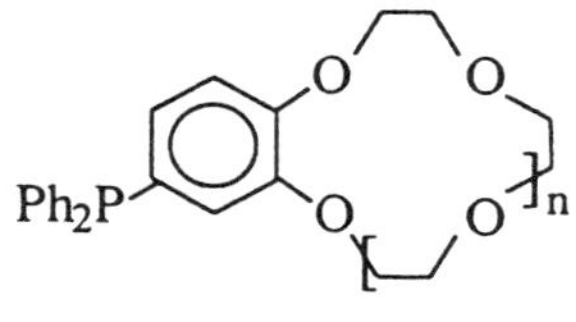

n=1, 2, 3, 4

80

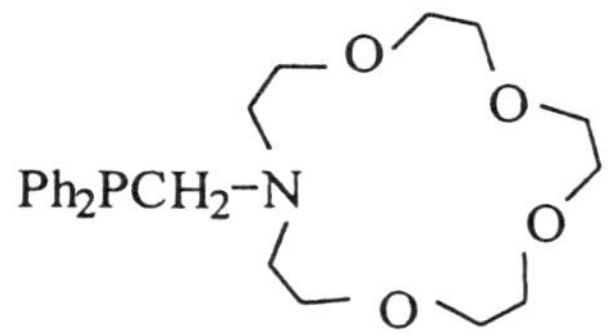

80a

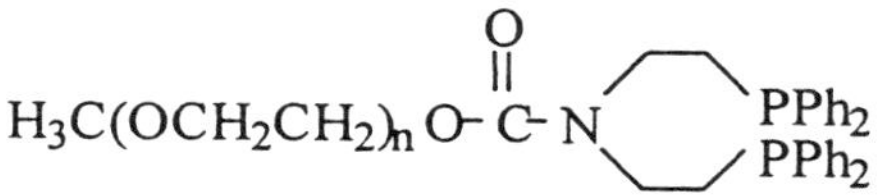

n = 12, 16, 110

81

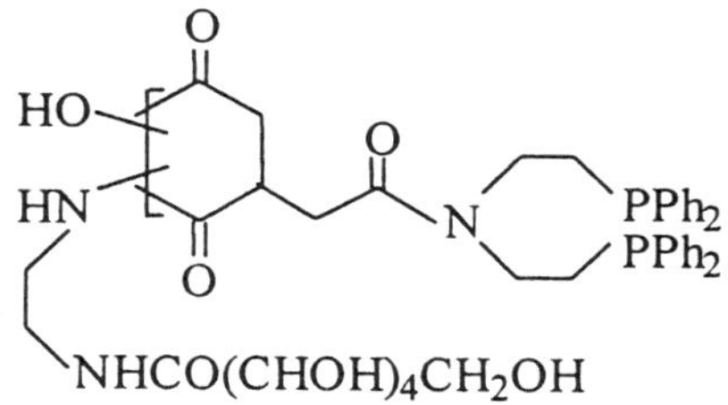

82

---

[a] **References**: **57**[31, 48, 51, 53, 61, 184, 192-196, 198]; **58**[53, 62, 66, 162, 194, 197, 212, 213]; **59**[41, 194, 198, 228]; **60**[194]; **61**[41, 194, 198]; **62**[41, 199]; **63**[41]; **64**[41]; **65**[164, 165, 200, 201, 204, 205, 238]; **66**[201, 205]; **67**[31, 50, 89, 200-203]; **68**[201-203]; **69**[206]; **70**[206]; **71**[51, 53, 61, 63, 66, 118, 123, 207]; **72**[61, 63, 66, 123, 208]; **73**[209]; **74**[209]; **75**[209]; **75a**[500]; **76**[55, 56, 214, 215, 259, 371]; **77**[55, 56, 123]; **78**[55, 56]; **79**[150, 151, 330]; **80**[53, 55, 56, 61, 123, 216]; **80a**[372]; **81**[61, 66, 68, 150, 151, 190]; **82**[62, 66, 150, 151]

Table 5 *Water soluble phosphines containing amino, ammonium, phosphonium, phosphonate and phosphate functionalities*[a]

$[R_2P(CHR')_nNMe_2R'']^+$ R=Ph, Cy ; R' =H, Me ; n= 2, 3 ; R" = H, Me, iPr **83**	$P[(CH_2)_nNR_2]_3$ n =2,3; R=Me, Et **84**	$P[(CH_2)_2NMe_2R]_3^+\ 3X^-$ R = H, Me **85**	$[(HOCH_2)_2P(CH_2)_2NMe_3]^+$ **86**
$[R_2P(CH_2)_2NMe_2R']^+$ R=R' = Me, Octyl, Dodecyl **87**	$[H_2P(CH_2)_nNR_2R']^+$ n = 2, 3, 6, 7, 8, 12 R=R' = Me, Et, Butyl **88**	$[Me_2P(CH_2)_2NMe_2R]^+$ R = Octyl, Dodecyl **89**	$\{H_2N[(CH_2)_2PPh_2]_2\}^+$ **90**
$Cy_nP[(CH_2)_2N^+Me_3]_{3-n}$ n =1, 2 ; Cy= Cyclohexyl **91**	$Ph_nP[(CH_2)_3NHC^+(NH_2)_2]_{3-n}$ n =1, 2 **92**	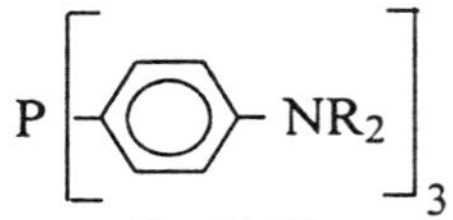 R = H, Me **93**	

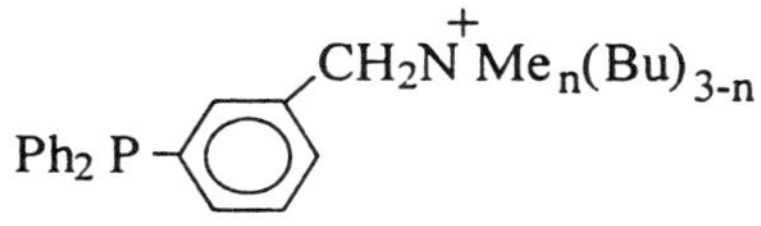

n=1, 2, 3

94

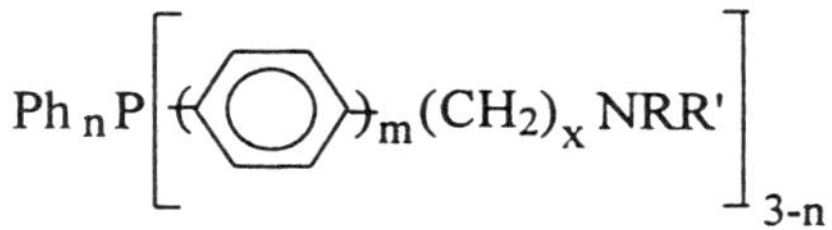

n=1, 2; m=0, 1; x=1, 2; R=Me, Et, Ph
R' = Me, Et, Ph

95

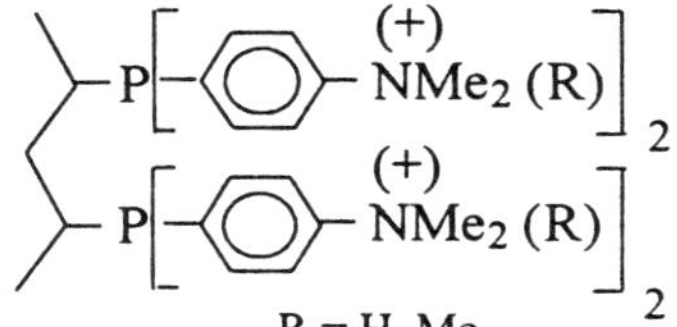

R = H, Me

96

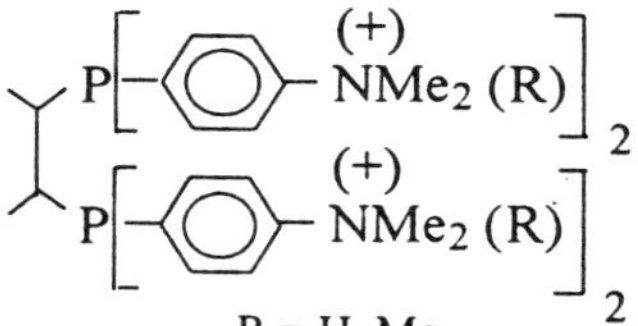

R = H, Me

97

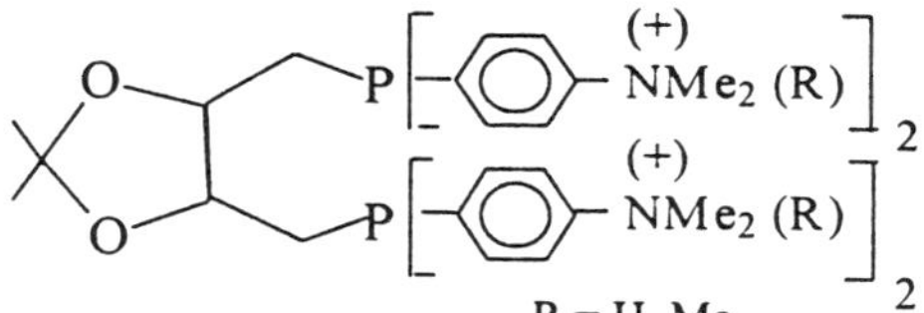

R = H, Me

98

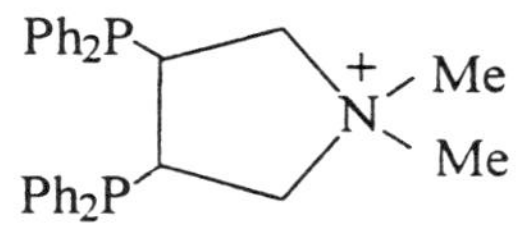

99

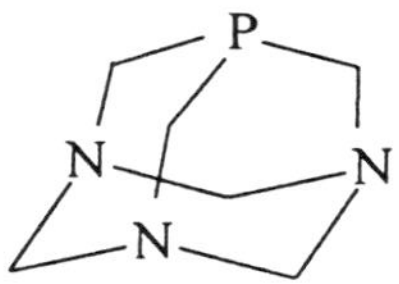

100

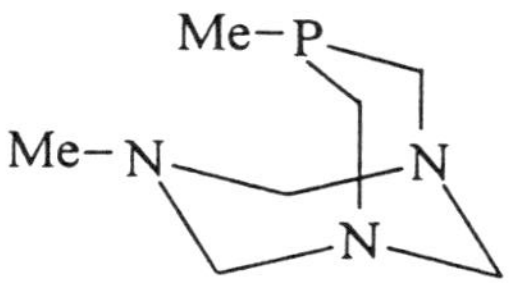

101

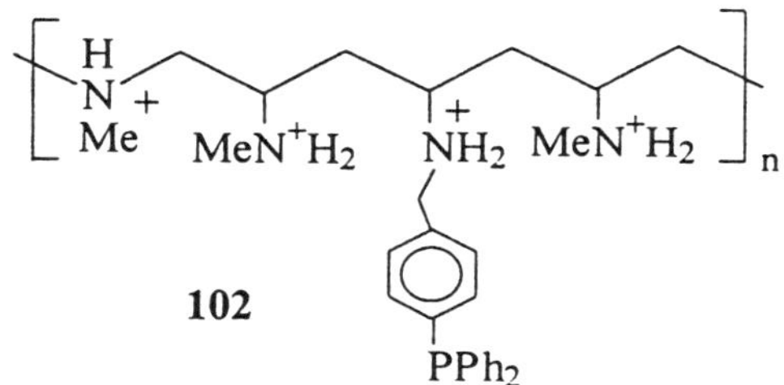

102

Table 5 *(contd)*

$[Ph_2P(CH_2)_nPMe_3]^+$
n=2, 3, 6, 10

103

$Ph_2P(CH_2)_2P(=O)(ONa)_2$

104

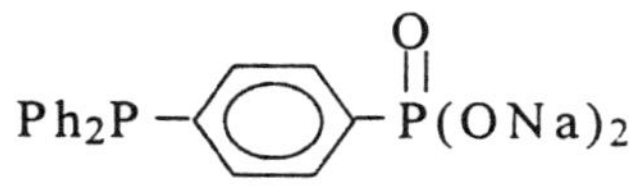

105

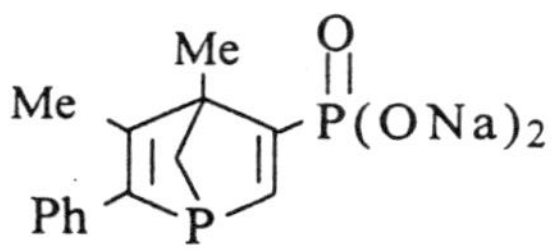

106

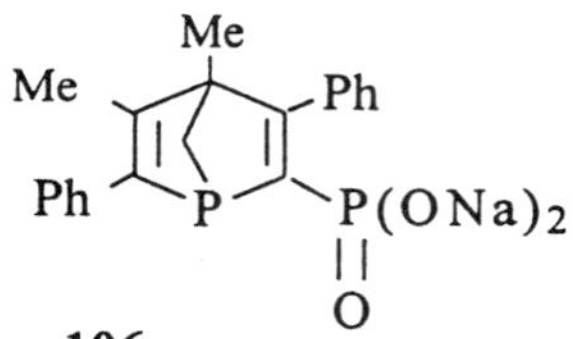

106a

$Ph_2P(CH_2)_{11}OP(=O)(ONa)_2$

107

---

[a] **References**: **83**[31, 37, 48, 51, 53, 60, 61, 63, 65, 66, 87, 123, 129, 217-220, 227, 230-232, 259]; **84**[227, 238, 330, 357, 369]; **85**[227, 231]; **86**[227]; **87**[227, 231]; **88**[227, 230]; **89**[231]; **90**[61, 87, 150, 151, 190, 231]; **91**[129]; **92**[221, 233]; **93**[53, 164, 234-237]; **94**[55, 288]; **95**[53, 62, 165, 221, 238, 320]; **96**[51, 53, 61, 227, 231, 239-243]; **97**[51, 53, 61, 239, 240-243]; **98**[51, 53, 61, 239, 240, 243, 259]; **99**[53, 66, 231]; **100**[51, 53, 221, 244-254, 261, 320]; **101**[221, 254]; **102**[188, 501]; **103**[53, 61, 221, 227, 230, 255-257, 259]; **104**[51, 128, 221, 258, 259]; **105**[221, 258, 259, 260]; **106**[260]; **107**[53, 262]

Table 6 *Miscellaneous water soluble ligands*[a]

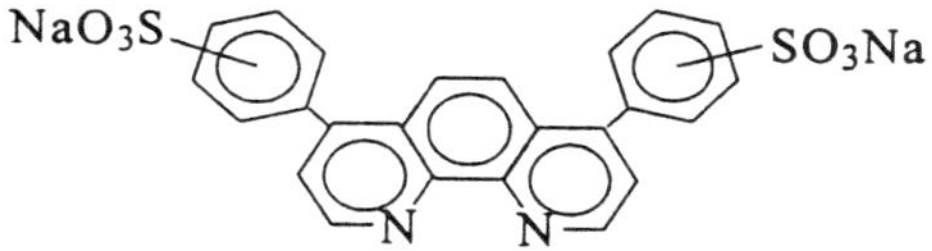

108

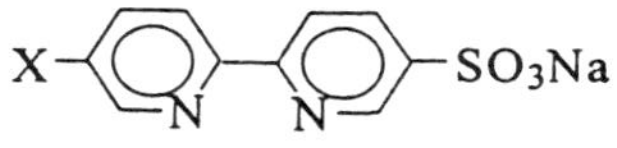

X = H, SO_3Na

109

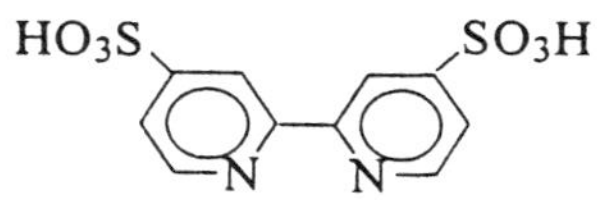

110

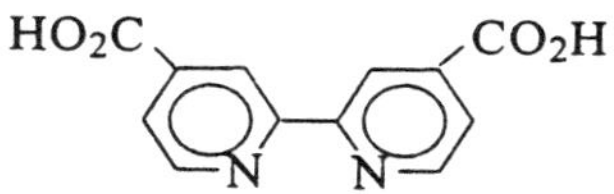

111

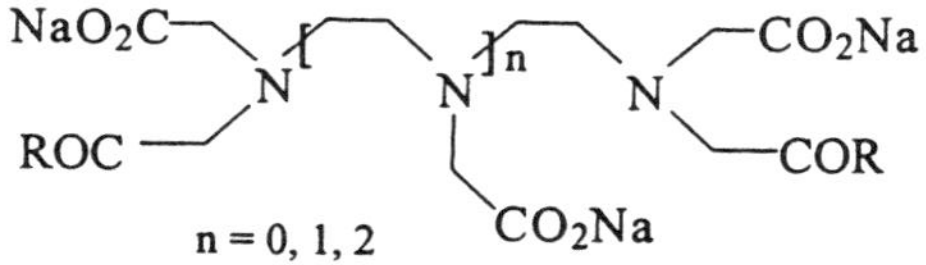

n = 0, 1, 2

R= ONa, $NHCH_3$, $NH(CH_2)_2CH_3$

112

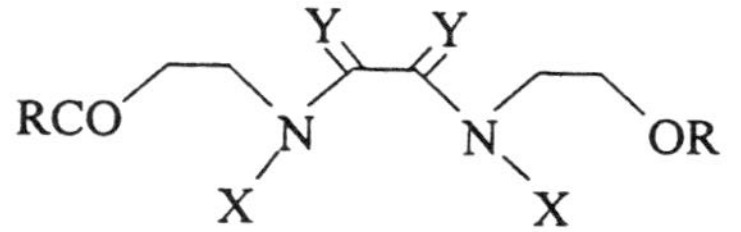

X=Me, $(CH_2)_2OH$, $(CH_2)_2OMe$

Y= 2H, O ; R=H, Me

113

Table 6 *(contd)*

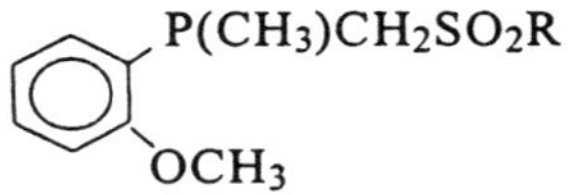

R= Me, NMe_2

114

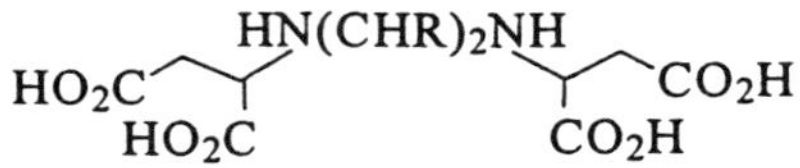

R=H, Ph

115

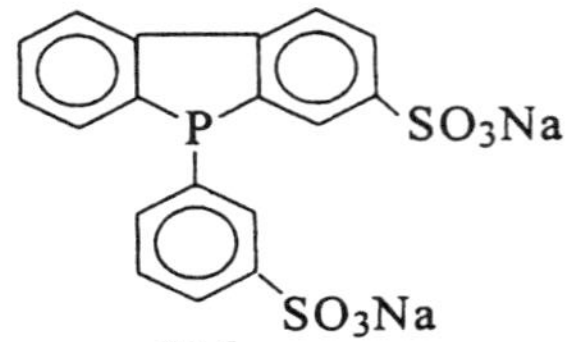

116

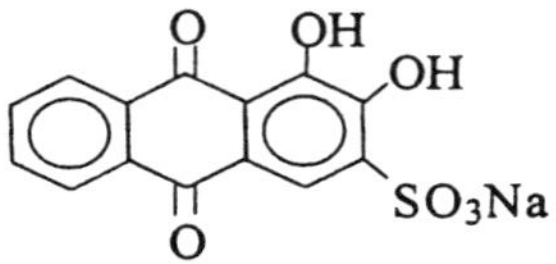

117

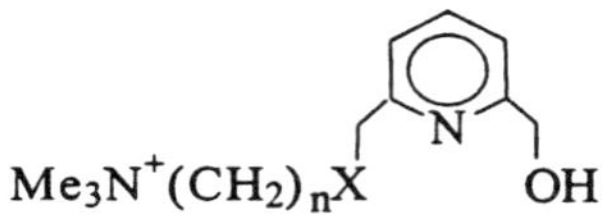

n=2, 10, 11, X= CONH, S

118

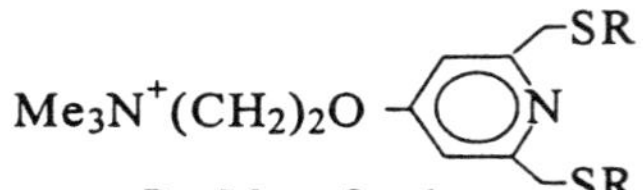

R = Me, n-Octyl

119

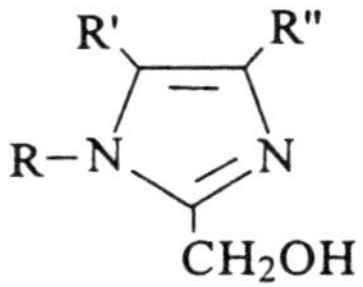

R = $C_{16}H_{33}CH(OSO_3Na)CH_2$, $C_{12}H_{25}$

R' = H , CH_2OH ; R" = H , $CH_2S(CH_2)_2SO_3Na$

120

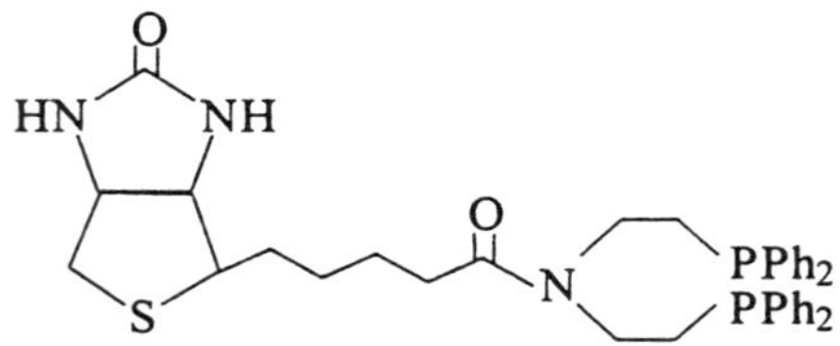

121

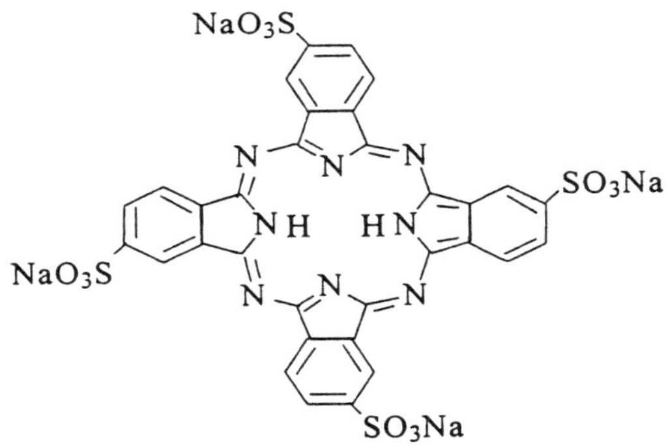

122

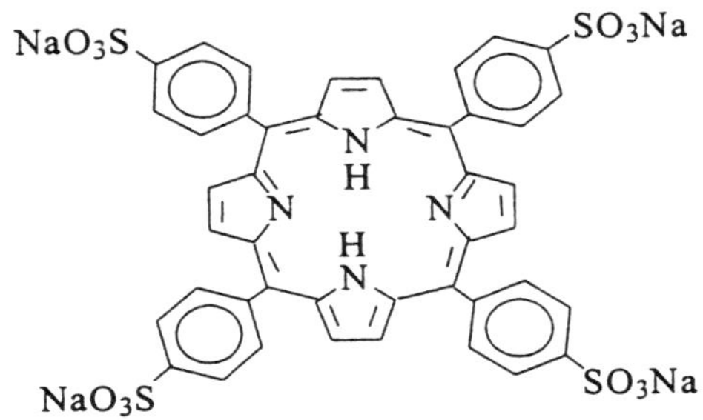

123

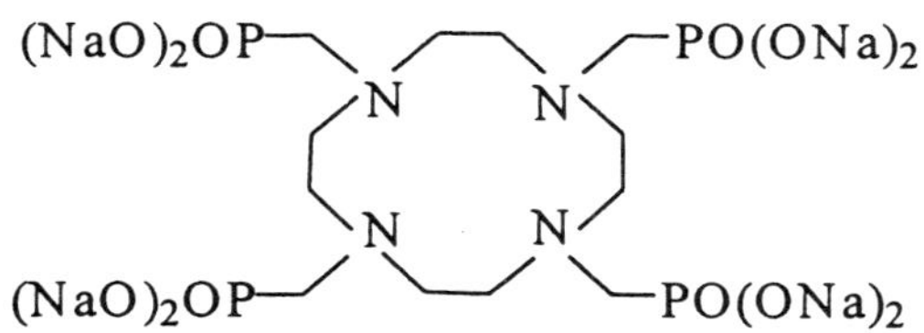

124

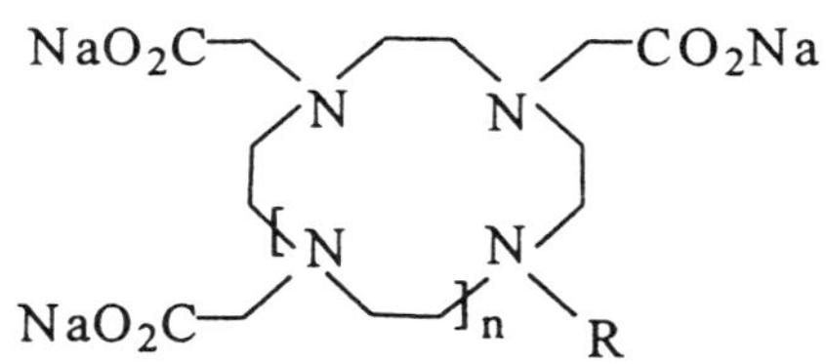

n = 0, 1

R = CH_2CO_2Na, $CH_2CH(OH)CH_3$

125

Table 6 *(contd)*

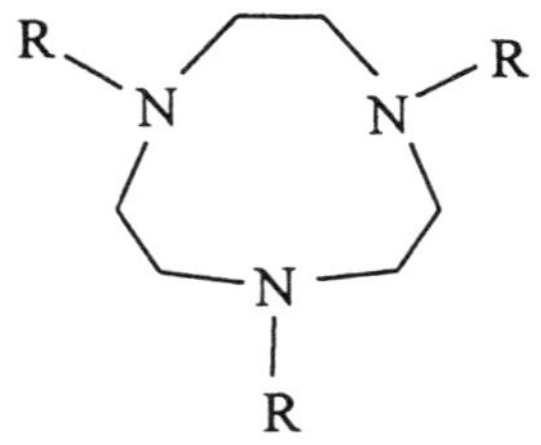

R = CH_3, $(CH_2)_2OH$, $(CH_2)_2CO_2H$

126

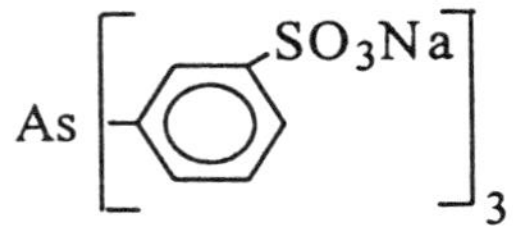

127

$Ph_nP[(CH_2)_2SCH_2R]_{3-n}$

n = 0, 1, 2

R = CH_2SO_3Na, $(CH_2)_2SO_3Na$, CH_2NH_2, $CH(NH_2)CO_2Et$

128

$Ph_2PCH{=}CHS(CH_2)_3SO_3Na$

129

$Ph_2PCH_2CH[S(CH_2)_3SO_3Na]_2$

130

---

[a] **References**: **108**[63, 263-266]; **109**[62, 63, 267-269]; **110**[62, 269]; **111**[62, 269, 270]; **112**[62, 271, 292, 295]; **113**[272]; **114**[123, 273]; **115**[292, 293, 300, 301]; **116**[82, 221]; **117**[274, 275]; **118**[276, 277]; **119**[276]; **120**[278, 279]; **121**[53, 62, 68, 123, 151, 190]; **122**[280, 281, 317-319, 368]; **123**[290-292, 296-299]; **124**[292, 294, 302-304]; **125**[292, 293, 294, 295]; **126**[292-294, 332]; **127**[98, 99]; **128**[333]; **129**[333]; **130**[333]

A few years ago, a new class of ligands namely the sulfonated phosphites (for examples see Table 7, **132**, **133**) was described.[283-287] They show remarkable stabilities in water compared to conventional phosphites such as $P(OPh)_3$ and rhodium catalysts modified with **132** exhibited much higher catalytic activities in the hydroformylation of 1-tetradecene than conventional $Rh/P(OPh)_3$ or Ph/PPh_3 catalysts even at lower reaction temperatures.[285,286] Sulfonated phosphite ligands may play a role in the emerging field of biphasic catalysis in ionic liquids[15-22] or in combination with membrane separation of the metal complexes of these bulky ligands.

3 Olefin Hydroformylation

The hydroformylation reaction (oxo synthesis, Roelen reaction) is the catalytic conversion of olefins with synthesis gas to aldehydes, equation (1), discovered by Roelen[310-313] at Ruhrchemie AG in 1938 during his studies on the Fischer-Tropsch (FT) synthesis using a *heterogeneous* FT catalyst consisting of cobalt, thorium oxide, magnesium oxide and kieselgur. Subsequently, Roelen[314] as well as Adkins and Krsek[315,316] recognized the homogeneous catalytic nature of the hydroformylation reaction.

$$R\text{-}CH{=}CH_2 + CO/H_2 \rightarrow R\text{-}CH_2\text{-}CH_2\text{-}CHO + R\text{-}CH(CH_3)\text{-}CHO \qquad (1)$$

Nowadays, hydroformylation is one of the most important large-scale applications of homogeneous catalysis and the world market for oxo-products was estimated in 1995 at over 6.6 million t/a.[48] The conversion of propene to n- and i-butyraldehyde[356] is commercially important as n-butyraldehyde is converted, by aldol condensation and hydrogenation to the plasticizer 2-ethylhexanol. The i-butyraldehyde is generally regarded as a waste product and optimization of the process has been primarily aimed at increasing the selectivity towards the linear product (n-aldehyde). We note, however that i-butyraldehyde can be used for the production of methyl methacrylate or neopentyl glycol. Table 8 compares the various propene hydroformylation processes[321-323] employing cobalt and rhodium catalysts.

Cobalt was the first catalyst used in commercial applications of the oxo-synthesis. In order to stabilize the $HCo(CO)_4$ catalyst, high pressures are necessary with a maximum n/i ratio of 80/20. In the Shell process,[324,325,393] cobalt catalysts modified with alkylphosphines e.g. $HCo(CO)_3P(C_4H_9)_3$ are more selective towards linear products but exhibit high hydrogenation activity and are therefore mainly used for the direct synthesis of long chain alcohols.

Following the discovery of the $RhH(CO)(PPh_3)_3$ catalyst by Wilkinson,[326] Celanese[327,328] and Union Carbide[329] independently developed a rhodium-based oxo process operating at mild reaction conditions and exhibiting high n/i ratios. The main limitation of the so-called Low Pressure Oxo (LPO) process is that it is applicable only to lower olefins (up to 1-pentene) because the catalyst is separated from the aldehyde products by distillation. In principle, aldehydes of up to 10 carbon atoms chain length can be distilled, but the temperatures required for the

Table 7 *Other sulfonated ligands*[a]

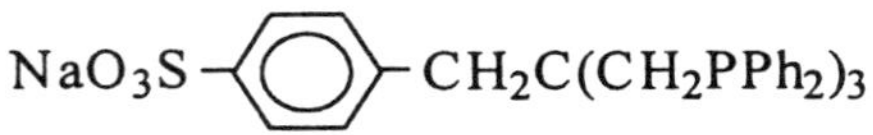

131

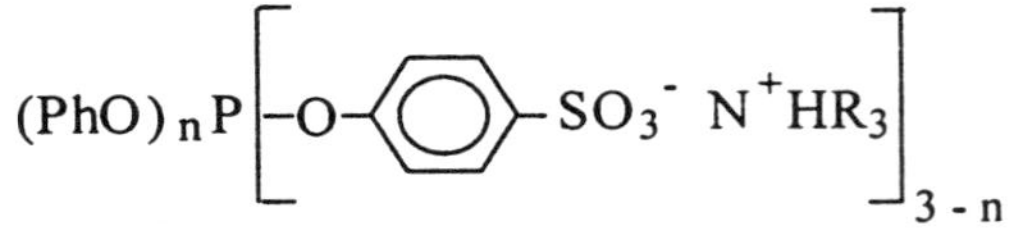

n = 0, 1, 2 ; R = C_8H_{17}

132

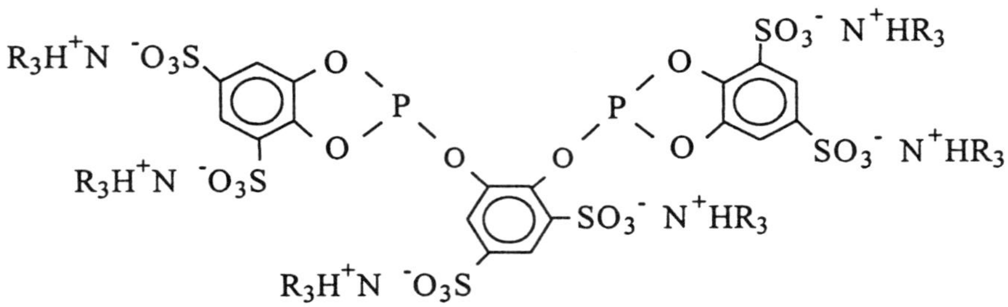

133

[a] **References: 131**[282]; **132**[61, 283-287]; **133**[283-286]

Table 8 *Comparison of various industrial propene hydroformylation processes*

Process	*Classical Ruhrchemie BASF ICI*	*Shell*	*Ruhrchemie*	*LPO-Process*[a]	*Ruhrchemie/Rhône-Poulenc (RCH/RP) Process*
Catalyst	$HCo(CO)_4$	$HCo(CO)_3PR_3$	$HRh(CO)_4$	$RhH(CO)(PPh_3)_3$	$RhH(CO)(tppts)_3$
Temperature [°C]	110–180	160–200	100–140	85–115	50–130
Pressure [bar]	200–350	50–100	200–300	15–20	10–100
Selectivity to aldehydes	medium	low	high	high	high
Selectivity to alcohols	medium	high	low	low	low
n/i ratio	80/20	88/12	50/50	92/8	95/5[b]
Hydrogenation	medium	high	low	low	low
Sensitivity towards catalysts poisons	low	low	high	high	no sensitivity
Separation Catalyst/Product	chemical	thermal	chemical	thermal	decantation

[a] Low Pressure Oxo (LPO) Process of Wilkinson/Union Carbide/Johnson Matthey/Davy McKee; [b] In the mean time n/i = 99/1.[344]

separation cause extensive catalyst degradation and formation of heavy ends. Even in the hydroformylation of propene heavy ends formed during the aldehyde distillation and accumulation of sulfur-containing compounds in the catalyst phase eventually leads to deactivation of the catalyst.

A breakthrough in the hydroformylation of propene was achieved following the synthesis of the water soluble ligand tppts for the preparation of the $RhH(CO)(tppts)_3$ catalyst[345] which formed the basis for the development of the Ruhrchemie/Rhône-Poulenc two phase process. This process operates under mild reaction conditions giving excellent n/i ratios and easy separation of products from the catalyst by decantation with virtually no catalyst leaching.

In 1995, nearly 80% of all oxo products and over 90% of the propene hydroformylation products were produced using rhodium-based catalysts.[48]

3.1 Hydroformylation of Propene: the Ruhrchemie/Rhône-Poulenc (RCH/RP) Process – The industrial process for the hydroformylation of propene in a two phase system was developed by Ruhrchemie AG after a period of successful scale up tests in pilot plants.[31,34,35-38,42,48-50,52,57,61,64,67,70-72,74,132-134,224,307,322,323,336-344,348,356,380-385,407a,415,418,422,424]

The first commercial plant, with a capacity of 100 000 t/a n-butyraldehyde, went on stream in July 1984.[323] Today the capacity is 300 000 t/a butyraldehydes (debottlenecked to 400 000 t/a) and a cumulative total of more than 2 million tons has been produced since the introduction of the process.[37,38,350]

Typical reaction conditions of the RCH/RP process[37,38,336] are T=120°C; P=50 bar; $CO/H_2=1.01$; P/Rh=50-100; aqueous/organic phase=6; concentration of rhodium: 10-1000 ppm; pH of the aqueous phase 4-10. The $RhH(CO)(tppts)_3$ catalyst[345] is prepared *in situ* from rhodium-2-ethylhexanoate, for example, by addition of tppts in water. The process engineering is enormously simplified in comparison to classical oxo plants. Figure 1 shows the flowsheet of the RCH/RP oxo process.[38,424]

The selectivity toward butyraldehydes is 99% with an n/i ratio of 95/5. Competing hydrogenation of the aldehyde product, the formation of heavy ends (aldol condensation products) and formates is negligible. Rhodium carry-over from the aqueous to the organic phase is in the ppb range. The $RhH(CO)(tppts)_3$ catalyst in the two phase system is not sensitive to catalyst poisons because they are contained in the organic phase. The manufacture and handling of tppts has not presented any problems. Water soluble decomposition products which accumulate in long term over the catalyst's life are separated, after the decantation of the aqueous catalyst solution in the phase separator 2, by a combination of membrane separation and an extraction/re-extraction process affording a rhodium and ligand recovery of 99.95% and 98%, respectively.[307,349]

In short the RCH/RP process has significant economic and environmental advantages: easy and complete catalyst recovery, high catalyst activity and selectivity, simpler operation, recovery of the exothermic heat of the oxo reaction, virtual elimination of plant emissions and the avoidance of organic solvent. The costs of the RCH/RP process are estimated to be 10% lower than other oxo processes.[38]

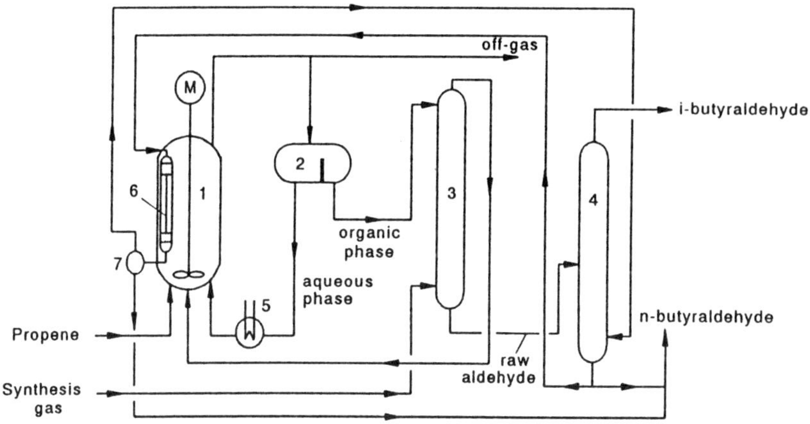

Figure 1 *Flowsheet of the RCH/RP hydroformylation process:*[38] *1 Continuous flow stirred tank reactor,*[424] *2 Phase separator, 3 Stripping column, 4 Distillation column, 5 Heat exchanger, 6 Falling film evaporator, 7 Liquid/vapor separator.*

Based on the success of this process much effort is being devoted to developing new water soluble ligands in order to increase the efficiency of the rhodium catalyst for the hydroformylation of propene even further. Examples of novel ligands include the sulfonated phosphines **24** (norbos), **25** (binas) and **27** (bisbis, when X=SO_3Na, Z=H, Y=C_6H_4-*m*-SO_3Na, x=1, n=0,1) depicted in Table 2. Rhodium catalysts modified with bisbis and norbos exhibit an activity of 5.6 and 7.4 times that of tppts, respectively. The n/i ratio achieved in the hydroformylation of propene is 97/3 and 81/19 with Rh/bisbis and Rh/norbos, respectively, compared to 94/6 with Rh/tppts.[31,35,37,38,49,61,132-134,221,224,344,374] Rh/binas catalysts exhibited an even higher catalytic activity in the biphasic hydroformylation of propene, i.e. twelve times that of Rh/tppts, with a n/i ratio of 98/2.[31,37,49,132,133,221,224,344] Rhodium catalysts modified with the chelate ligands binas and norbos exhibit higher catalytic activity in the biphasic hydroformylation at substantially lower phosphorus/rhodium ratios than with tppts (*e.g.* binas:6.8/1; norbos:13.5/1; tppts:80/1).[132] Rh/binas complexes have been tested as catalysts in the propene hydroformylation in lab and in pilot plant scale during a period of 2 months without any decrease in catalytic activity being observed.[344]

3.2 Hydroformylation of Other Lower Olefins and Dienes – Lower olefins such as 1-butene or 1,3-butadiene are hydroformylated with acceptable rates using Rh/tppts catalysts according to the RCH/RP process. Hoechst AG Werk Ruhrchemie has developed an attractive new process[350] for the hydroformylation of raffinate II, a mixture of 1-butene, *cis*- and *trans*-butene-2 and n- and i-butane derived from the C_4 stream of naphtha crackers (after removal of 1,3-butadiene and i-butene).[351,354,355] A raffinate II mixture rich in 1-butene is hydroformylated to n-valeraldehyde and 2-methylbutanal (Figure 2) with an n/i ratio of 95/5 using Rh/tppts catalysts at 120-130°C, 40-60 bar and a rhodium concentration of 150-500 ppm in an aqueous/organic two phase system.[350]

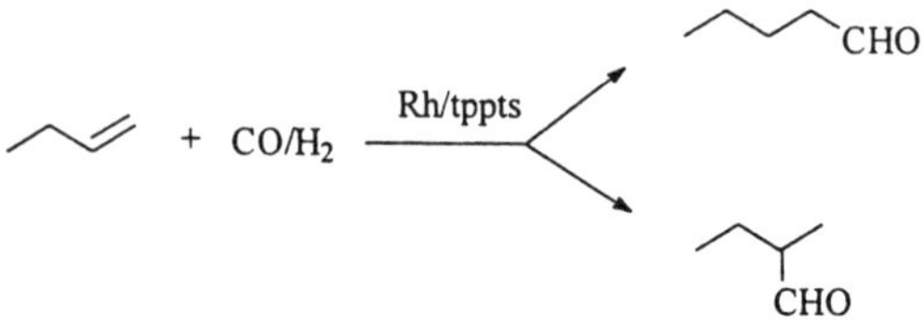

Figure 2 *Rh/tppts-catalysed hydroformylation of 1-butene.*

The advantages of this new process are the high n/i ratio and the low formation of heavy ends (1.5%) and alcohols (1%) without any formation of formates. In the original cobalt-based high-pressure process the n/i ratio was 67/33, with substantial formation of heavy ends (6.7%), alcohols (6.6%) and pentylformates (4.2%).[350] The n-valeraldehyde product is oxidized with molecular oxygen to n-valeric acid, the trimethylolpropane, pentaerythritol, or dipentaerythritol esters of which are used as lubricants.[350]

DSM[370] has patented platinum systems based upon tetrasulfonated bidentate water soluble ligand **29** (Table 2; x=4, m=0, n=0) as catalysts for the hydroformylation of a mixture 1-butene (45%) and 2-butene (22%) with 33% butane at 100°C and 80 bar CO/H_2 in an aqueous/methanol (300/32), CF_3SO_3H acidic medium. The olefin conversion was 86% and the selectivity to the aldehydes 95% (n/i ratio of 2.8) together with small amounts of aldolcondensation products and acids. The products were isolated from the aqueous catalyst mixture leaving the reaction zone by extraction with ether and the aqueous phase recycled to the reactor.

1,3-Butadiene was hydroformylated using Rh/tppts catalysts at P/Rh molar ratios of 15-80, 100-140°C, 55-200 bar CO/H_2, 410 ppm of rhodium, a pH of 9, in an aqueous/organic two phase system and subsequently hydrogenated using Pt/C (10 wt.% Pt) at 140°C and 200 bar H_2 to give 42-67% n-pentanol and 9-23% 2-propylheptanol together with cyclopentanemethanol (1-12%) and 2-methylbutanol (2-4%).[352] The formation of the aldolcondensation product, 2-propylheptanal, decreased with decreasing pH and increasing temperature.

3.3 Hydroformylation of Mid Range Olefins – Rhodium/tppts catalysts exhibit low catalytic activity in the hydroformylation of mid range olefins (C_5-C_8) in a two phase system due to the much lower solubility of such olefins in water. In the Rh/tppts catalysed biphasic hydroformylation of 1-hexene, for example the conversion is only 11-22% with a n/i ratio of aldehydes of 98/2.[353,373] The rate of 1-hexene hydroformylation catalysed by Rh/tppts increased by a factor 2.3 when subjected to ultrasound (35 kHz) and high stirring rates.[360,361]

Chaudhari *et al.*[363,364] reported that the rate of biphasic hydroformylation of 1-octene (water solubility 0.024 mol/m^3 compared to 7.84 mol/m^3 for propene at 25°C and 1 atm[50,477]) using $RhH(CO)(tppts)_3$ was increased by a factor 10 to 50 when the "promotor ligand" PPh_3 was added. Carry-over of rhodium from the aqueous to the organic phase was found to be less than 5 ppm.[363] These results were rationalised by assuming that the *in situ* formation of mixed ligand complexes, $RhH(CO)(tppts)_{3-x}(PPh_3)_x$ promoted catalysis at the interface between the organic and aqueous phase.

A relatively weak rate enhancement was observed in the biphasic hydroformylation of 1-octene using Rh/tppts catalysts in the presence of cosolvents such as ethanol to enhance the solubility of the olefin in the aqueous phase and with addition of buffers ($Na_2CO_3/NaHCO_3$) to eliminate side reactions such as the formation of acetals.[31,365,366] Similarly, addition of ethanol improved the yields in the hydroformylation of 1-octene catalysed by $Rh_2(\mu\text{-S-}^t\text{Bu})_2(CO)_2(tppts)_2$ species in an aqueous/organic two phase system.[367] However, the high selectivity to linear aldehyde observed for neat olefin in the biphasic system (97%) decreased (to 83%) in the presence of the cosolvent.[367]

Hanson *et al.*[79] investigated the effect of ionic strength of the aqueous phase on the biphasic $RhH(CO)(tppts)_3$-catalysed hydroformylation of 1-hexene. They showed that the addition of monovalent "spectator cations" such as Li^+, Na^+, Cs^+ resulted in a decrease in rate and an increase in n/i ratio with increasing ionic strength.[79] These results were rationalised by assuming that high ionic strength in the aqueous solution stabilizes $RhH(CO)(tppts)_2$ species rather than RhH(CO)(tppts) via anion/cation interactions and hydrogen bonding in the aqueous medium.[79,81,345]

Replacement of tppts by the fluoro substituted sulfonated ligand **4** [Table 2; 94% (n=1) and 6% (n=0)] in the rhodium-catalysed hydroformylation of 1-hexene in a two phase system increased the selectivity to linear aldehyde n-heptanal from 86% to 93% at the low P/Rh molar ratio of 7.5/1.[75,76] The Rh/**4** catalyst was quantitatively recovered after the reaction by simple decantation.[75,76] The moderate increase of the n/i ratio is of interest when one considers that ligand **4** is mainly present as the disulfonated species (94%) compared to the trisulfonated compound tppts and that tris(4-fluorophenyl)phosphine is less basic (pK_a=1.97) than triphenylphosphine (pK_a=2.73).[376] In rhodium-catalysed hydroformylation reactions in organic solvents it is known that electron withdrawing substituents, which increase the π-acidity of the ligand, give rise to an increase in the n/i ratio.[377-379]

Table 9 summarizes further biphasic hydroformylation reactions of various mid range terminal and internal olefins such as 1-hexene, 1-octene and 2-hexene catalysed by different water soluble systems.

Kalck *et al.*[96,359,362] studied the biphasic hydroformylation of 1-hexene using water soluble dinuclear species such as $Rh_2(m\text{-S-}^t\text{Bu})_2(CO)_2(tppts)_2$ and CO/H_2O where H_2O acts both as a solvent and hydrogen source according to the water gas shift reaction (Equation 2). The turnover frequency (TOF) obtained was 40 h^{-1} and the n/i ratio of the aldehyde 96/4.[96] Using $RhH(CO)(tppts)_3$ catalysts lower rates were obtained under the same conditions.[96]

Table 9 *Hydroformylation of 1-hexene, 1-octene and 2-hexene in a two phase system*

Olefin	*Catalyst*	*Temp [°C]*	*Press. [atm]*	*Yield[a]*	*n/i-ratio*	*Leaching of metal*	*Ref.*
1-Hexene	Rh/tppms	>70	f	f	f	yes	29
1-Hexene	Ru/tppms	90	60	30	3	f	29
1-Hexene	Pt/tppms	>90	f	f	f	decomp.	29
1-Hexene	Rh/**83**(amphos)[b]	90	40	90	4	yes	217,358
1-Octene	Rh/**29**(dppets)[c]	120	14	25	3.2	f	139
2-Hexene	Pt/**30**(m=0,n=0)[d]	100	80	15[e]	2.3	f	370

[a] Aldehydes; [b] amphos: **83** when R=Ph, R′=H, n=2, R″=Me (Table 5); [c] dppets: **29** when x=2, n=0, m=0 (Table 2); [d] see Table 2; [e] heptanoic acids (7%); f: not reported

$$CO + H_2O \rightarrow CO_2 + H_2 \qquad (2)$$

Water soluble Rh/tppts and Rh/tppms complexes dissolved in nonaqueous media such as the ionic liquids, 1-ethyl-3-methylimidazolium or 1-n-butyl-3-methylimidazolium salt have also been used as catalysts in the hydroformylation of 1-pentene employing a two phase system.[15,16] The yields obtained were 16-33% (TOF=59-103 h^{-1}) without any leaching of the rhodium from the ionic liquid to the aldehydes/feedstock phase. Rh/PPh_3 catalysts exhibited higher rates (TOF=333 h^{-1}) for the same biphasic reaction albeit with leaching of rhodium due to the uncharged nature of the catalytic system.[15]

3.4 Hydroformylation of Higher Olefins – The rates in the biphasic Rh/tppts-catalysed hydroformylation decrease dramatically with increasing molecular weight of the olefin. For example, 1-hexene is hydroformylated with conversions up to 22% using the Rh/tppts catalyst in a two phase system, whereas under the same conditions the practically water insoluble 1-tetradecene gives only traces of C_{15} aldehydes.[115,373] On the other hand, rhodium catalysts modified with the potassium salt of tppms afforded aldehyde yields up to 66% in the hydroformylation of higher olefins such as 1-dodecene in an aqueous/organic two phase system.[103] However, no data are available with regard to leaching of rhodium from the aqueous to the organic phase.[103]

3.4.1 Catalysis in Micellar Microheterogeneous Systems – A particularly elegant approach to circumventing the solubility problem in the biphasic hydroformylation of higher olefins is to use tenside phosphines to create micelles. To our knowledge, the water soluble tenside phosphine **15** (Table 2; n=0,3,5,7,9,11) is the first system which was specifically designed to combine the functions of a ligand and a surfactant for use in transition metal catalysed micellar reactions of water insoluble starting materials in aqueous media.[115,116] The rates in the Rh/**15**-catalysed hydroformylation of 1-tetradecene (olefin/Rh=1275, Rh concentration= 200 ppm) increased with increasing length of the hydrophobic hydrocarbon chain of the ligand **15** up to a maximum for C_8 (n=5) giving a conversion of 79% within

3 h reaction time (TOF=340 h^{-1}).[115,116] In sharp contrast, Rh/tppts under the same biphasic conditions gave only traces of C_{15}-aldehydes.[115,116] A further increase in chain length resulted in a decrease in conversion (C_{10} and C_{12} gave 72% and 39%, respectively). The Rh/**15** catalyst was quantitatively recovered by simple phase separation after the reaction when the ligand **15** has a high *Hydrophilic-Lipophilic Balance* (HLB) value[387,388,389,402,403,499a] namely a shorter hydrophobic alkyl chain in the range C_3-C_{10} (n=0-7). In contrast, at lower HLB i.e. longer chains (C_{12}- C_{14} ; n=9,11) stable emulsions were formed which did not break in 12 months at room temperature. The concentration of the surfactant ligand **15** (n=5) in the aqueous hydroformylation phase was 8.4×10^{-2} mol/l[116] and is above the *critical micelle concentration* (CMC) which is for typical surfactants usually between 10^{-4} and 10^{-2} mol/l.[386,389,390]

In order to explain that the catalytic activity goes through a maximum as a function of the hydrocarbon chain length of the ligand **15** the simplified model of a two-dimensional ionic spherical micelle[386,389-392,407b,499b], depicted in Figure 3, was proposed.[116,169]

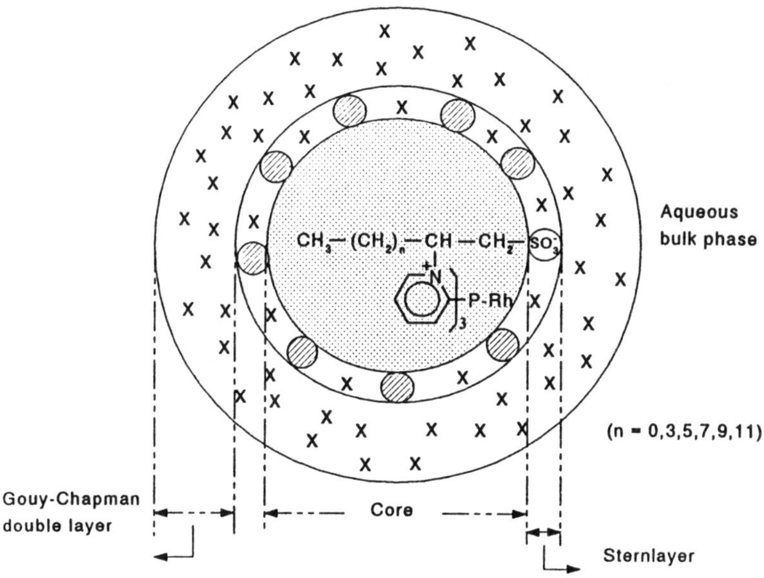

Figure 3 *Schematic two-dimensional simplified representation of a proposed spherical zwitterionic micelle containing the Rh/15 catalyst. The hydrocarbon chains of the tenside phosophine (-CH_2-CH-(CH_2)$_n$-CH_3), the hydrophilic head group (-SO_3^-), the counter ions (Na^+ and OH^-, depicted as X) and the solubilised 1-tetradecene (dotted part, core of the micelle) are schematically indicated to denote their relative locations and not their configuration, number, distribution or relationship to the molecular sizes.*

The *core* of the micelle is composed of the hydrophobic chain of the surfactant phosphine **15** where the higher olefin 1-tetradecene is solubilised. Surrounding the core is the *Stern layer* where the charged head groups (SO_3^-) of the surfactant **15** are located together with the small counter ions (Na^+) of the ionic micelle. The rhodium atom of the catalyst is probably located on the polarity gradient between the *Stern layer* and the *core* of the micelle.[169,479] Therefore the situation of the rhodium in the micelle should be dependent on the HLB value of the ligand **15**, *i.e.* on the length of the hydrocarbon chain in **15**; maximum efficiency being observed when n=5. In contrast, when n<5 the ligand **15** is probably too short to solubilise the olefin efficiently and when n>5 the Rh^+ is probably located far away from the polarity gradient between the *Stern layer* and the *core* of the micelle which gives rise to a drop in the catalytic activity.[115,116]

Alternatively, the Rh/**15**-catalysed hydroformylation of 1-tetradecene may proceed in "wet micelles"[478] (which are water-permeated micelles) or in reversed micelles[155,389,406] with solubilisation of an aqueous microdroplet of the Rh/**15** catalyst in the organic bulk phase [cf. Figure 4 (A)].

Several writers[53,169] have stressed the point that micellar catalysis has analogies with heterogeneous catalysis on solid surfaces because the solubilisation of substrate in the *core* of a micelle containing catalytically active sites is conceptually related to adsorption on surfaces and the number of micelles to surface area.

Hanson *et al.*[78,81,117,118,121,122,149] synthesized the tenside phosphines **16**, **17**, **19** and **34** (Table 2) and investigated the aggregation of the systems **16** (n=3,6) and **17**.[71,118] Hydroformylation reactions of 1-octene catalysed by Rh/**16** (n=3,6) in an aqueous methanol solution gave C_9-aldehydes and nonanols (up to 10%).[117,118] The TOFs achieved with Rh/**16**(n=3) and Rh/**16**(n=6) were 335 h^{-1} and 360 h^{-1}, respectively, compared to 260 h^{-1} obtained with Rh/tppts catalysts.[117]

Rh/**19** (n=1,2,3,6) catalysts were used in the hydroformylation of 1-octene in the presence of conventional tensides such as the sodium salt of the dodecylbenzene sulfonic acid (0.5 wt.%) in an 1-octene/nonane/methanol/water (60/34/50/56) mixture.[121] With Rh/**19** (n=6) the TOF achieved was 28 $h.^{-1}$[121]

$Na_5[Co^+(CO)_3(\mathbf{19})_2]^{5-}$ was used as catalyst for the hydroformylation of 1-hexene and 1-octene in a two phase system without leaching of cobalt into the organic phase.[122] The products obtained were almost exclusively aldehydes (4-38%) and very little (0.4-3%) or no alcohol formation[122] in contrast with cobalt/phosphine catalysed hydroformylation in organic solvents which give alcohols. The n/i ratios of the aldehydes were low (1.1-2.5),[122] however, and never approached that expected for a phosphine modified cobalt catalyst in non-aqueous media[324,325,393] (see Table 8).

Rh/**76** (Table 4; n=1, x=0, R= Me, Bu), which should be able to induce micelle formation, were used as catalysts in the biphasic hydroformylation of 1-dodecene.[371] The conversion was 80%, the n/i ratio 60/40 with no carry-over of the rhodium catalyst into the organic phase.[371]

Replacement of the sodium cation of the tppts by an ammonium group such as $Me(CH_2)_{13}N^+Me_3$ also imparts surfactant properties to the tppts system.[287,353,394,395] Hydroformylation of 1-hexene using water soluble rhodium catalysts modified with such tenside ammonium salts of tppts afforded up to 55% C_7-aldehydes (n/i=94/6).[353,394,395] Similarly, Rh/tppts catalysts in the presence of tenside phosphonium salts e.g. $Me(CH_2)_{13}P^+Et_3$ gave 74% yield to aldehydes (TOF=186 h^{-1}) in the biphasic hydroformylation of 1-tetradecene.[416]

Rh/**41** (Table 3, n=2) or Rh/tppms in the presence of tenside ammonium salts such as $Me(CH_2)_{11}N^+Me_3\ Br^-$ exhibited high selectivities toward n-aldehydes (n/i=96/4) at relatively high conversions (73-78%) in the micellar hydroformylation of 1-dodecene and 1-hexadecene at a high tenside/Rh ratio of 20.[155,157-159] The leaching of rhodium to the organic phase was lower than 1 ppm in the more favorable cases. Interestingly, minimal leaching of rhodium and high n/i ratios were observed at low stirring rates.[155] The fact that the stirring rate has an influence on the selectivity of the hydroformylation reaction was rationalised by assuming that, at tenside concentrations higher than the CMC, reversed micelles[389,406] were formed with poor mixing which favors the formation of the linear aldehyde due to steric constraints imposed on the olefin substrate (see Figure 4).[155]

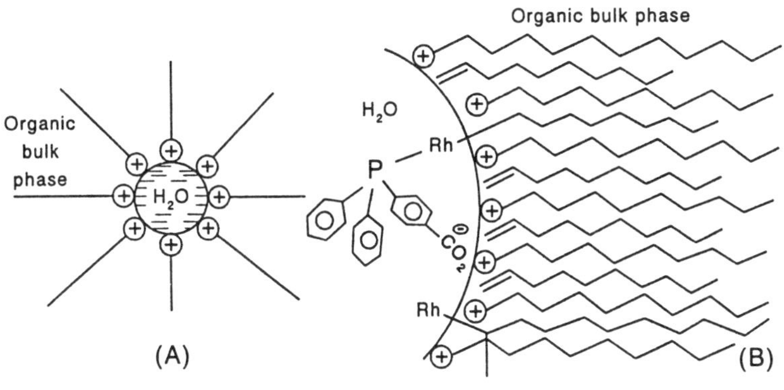

Figure 4 *(A) A spherical reversed micelle of a negatively charged micro droplet of water stabilised by cationic surfactant molecules. (B) Schematic representation of the steric interactions in the reversed micelle which favors the formation of linear alkyl rhodium intermediates.*

Rh/tppts catalysed hydroformylation of 1-dodecene was carried out in highly stable microemulsions generated by conventional surfactants e.g. sodium dodecyl sulfate and co-surfactants (butanol) which breaks after the reaction into two phases facilitating catalyst recovery.[414]

3.4.2 Thermoregulated Phase Transfer Catalysis – A conceptual advance in the field of biphasic hydroformylation of higher olefins is the use of rhodium catalysts generated from nonionic tenside phosphines, such as ethoxylated tris(4-

hydroxyphenyl)phosphines **67** (Table 4; n=0,1,2; m=3-25), which are organic soluble above a *critical temperature cloud point* (T_p) and on cooling to room temperature become water soluble, providing for facile catalyst separation by simple decantation.[200-203,396] The phenomenon of the inverse temperature dependent solubility in water at the T_p [well known for conventional nonionic ethoxylated tensides (T_p=50-100°C)[200,397-400]] can be explained in terms of a cleavage of the hydrogen bonds between the polyether chain and water by heating at $T \geq T_p$. Jin, Zheng and Fell[396] coined the term "*thermoregulated phase transfer catalysis*" to describe this kind of biphasic catalytic reactions where the site of catalysis is the organic bulk phase rather than the interface or the bulk aqueous phase at reaction temperatures higher than the T_p.[396] Conversions of 96% with 84% selectivity to aldehydes (TOF=33 h^{-1}) have been achieved in the hydroformylation of 1-dodecene using catalysts generated *in situ* from $RhCl_3{\cdot}3H_2O$ and the tenside ligand **67** (n=0, m=18), (T_p=95°C) at 100°C and 50 bar CO/H_2 in a two phase system.[396] After the reaction the organic phase was colourless indicating no leaching of rhodium and the catalyst was recycled four times without marked loss of activity and selectivity.[396,200] The less reactive olefin, cyclohexene, was similarly hydroformylated to afford cyclohexane aldehyde in 79% yield using Rh/**67** catalysts at 100°C within 90 min reaction time.[200] Rhodium catalysts generated from the surfactant **68** (Table 4) exhibited similar activity to the Rh/**67** system in the biphasic hydroformylation of 1-dodecene.[202,203]

Alternatively, Bianchini *et al.*[282] hydroformylated 1-hexene using rhodium catalysts generated from the sulfonated triphosphine **131** (Table 7) in water-methanol/isooctane or methanol/isooctane mixtures, which form a single homogeneous phase at 80°C but separate into two phases on cooling to room temperature thus providing the means for recovery of the catalyst.[282] In methanol/isooctane a mixture of alcohols and aldehydes (78/22) was obtained and at longer reaction times only alcohols whereas in aqueous methanol/isooctane only aldehydes (n/i=69/31) were obtained. However, the hydrogenation of 1-hexene nearly doubles in aqueous methanol/isooctane (46%) compared to methanol/isooctane (24%).

3.4.3 Monophasic Reaction with Biphasic Catalyst Separation and Recycling – An alternative concept for the hydroformylation of higher olefins is to perform the reaction in a monophasic organic system using rhodium catalysts modified with functionalised phosphines followed by separation and recycling of the catalyst in a two phase system. For example, Bahrmann *et al.*[287,404,405] used rhodium catalysts modified with water-insoluble triisooctylammonium salts of tppts in the hydroformylation of 1-decene or limonene in organic solvents. The conversion of 1-decene was 99% (n/i=65/35) at a rhodium concentration of only 44 ppm.[287] Limonene conversion was 98% at a[Rh]=132ppm.[287,404] The catalyst was recovered and recycled back to the reaction zone after addition of aqueous NaOH, phase separation, and subsequent re-extraction of the aqueous phase with triisooctylamine/toluene in the presence of sulfuric acid.[287]

Other examples of the application of this concept in the hydroformylation of mid range or higher olefins include: (i) rhodium catalysts generated from the amphiphilic ligands **41** (Table 3, n=2), **65** (Table 4; n=2, x=OH, m=0), **95** (Table 5; n=1, m=1, x=1, R=R′=Et; n=2, m=1, x=1, R=R′=Et; n=2, m=1, x=1, R=R′=Ph; n=1, m=1, x=1, R=Me, R′=Ph), $Ph_2P(4\text{-}C_5H_4N)$, $Ph_2P(3\text{-}C_5H_4N)$ and $PhP(3\text{-}C_5H_4N)_2$ in the hydroformylation of 1-octene in toluene;[165,238] (ii) cobalt catalysts modified with the amphiphilic ligand **84** (Table 5; n=3, R=Et) in the hydroformylation of 1-octene.[330,369]

Virtually quantitative conversions were observed in the hydroformylation of 1-tetradecene with rhodium complexes generated from the lithium salt of tppms or the lithium (sodium) salts of **21** (Table 2; R=Ph; n=3,4) and **22** (Table 2) in methanol as solvent.[127,334] Catalyst recycling involved evaporation of methanol and addition of water to form a two phase system, separation of the aqueous phase, evaporation to dryness and addition of MeOH.

Union Carbide[408-413] developed a new process for the hydroformylation of higher olefins using rhodium catalysts modified with the sodium salt of tppms which in the presence of certain solubilising agents, such as N-methylpyrrolidone (NMP) with trace amounts of water (<1 wt.%), or polyethylene glycols, $RO(CH_2CH_2O)_nH$, become soluble in non-polar higher olefin and yield a single phase system. After the reaction the solubilised Rh/tppms catalyst can be easily induced to separate into the polar catalyst and nonpolar aldehyde/feedstock phases thus providing an effective means for the recovery of the catalyst. In the preferred case of the NMP-solubilised system a sharp phase separation is induced by addition of excess of water or other polar compounds such as MeOH (Figure 5). A 86 day-long continuous olefin hydroformylation was achieved in a small experimental unit using Rh/tppms catalysts with high olefin efficiencies (>90%) under mild reaction conditions (110°C, 5-30 psia) producing C_7- through C_{15}-aldehydes.[409,411]

3.5 Hydroformylation of Functionalised Olefins – Introduction of polar substituents such as $\text{-}CO_2Me$ into the alkyl chain of α-olefins considerably extends the scope of Rh/tppts-catalysed hydroformylation in aqueous/organic two phase systems.[89,417]

Hydroformylation of methyl acrylate to the α-branched aldehyde (see Figure 6) catalysed by Rh/tppts complexes in an aqueous/organic two phase system proceeds with high reaction rates (TOF= 545 h^{-1}) which are 2.4 times higher than that exhibited by Rh/PPh_3 systems in completely organic media (TOF=225 h^{-1}).[423] This higher activity is unusual because an intrinsic limitation of biphasic catalysis is generally the significantly lower rates compared to analogous reactions in organic media.[70,345] The α-branched aldehyde product is a valuable fine chemical. For yields and selectivities obtained with various catalysts, see Figure 6.[130,423] Using Rh/tppms catalysts significant amounts of rhodium were carried-over to the organic phase.[130]

Methyl esters of ω-alkenecarboxylic acids were hydroformylated to give the corresponding aldehydes (Figure 7) using Rh/tppts catalysts at 120°C, 100 bar CO/H_2 and a rhodium concentration of 500 ppm in an aqueous/organic (2/1) two phase system.[417]

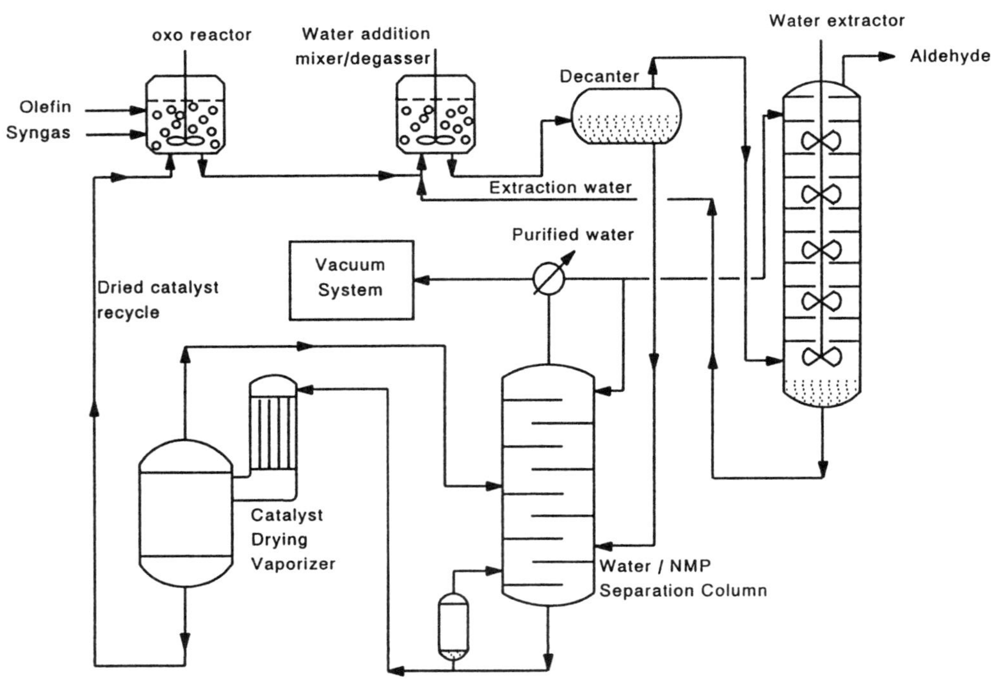

Figure 5 *Flow diagram of the Union Carbide Process for hydroformylation of higher olefins catalysed by Rh/tppms in a single phase with biphasic catalyst separation.*

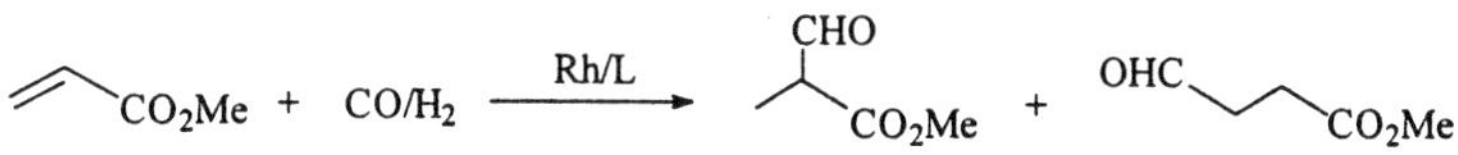

L	Yield (%)	i/n
PPh_3	95	>200
tppts	97	128
29 (x=4, m=n=0)	32	>200
tppms	83	63
56	68	25
23	60	22

Figure 6 *Rhodium-catalysed hydroformylation of methyl acrylate using various ligands.*[130,423]

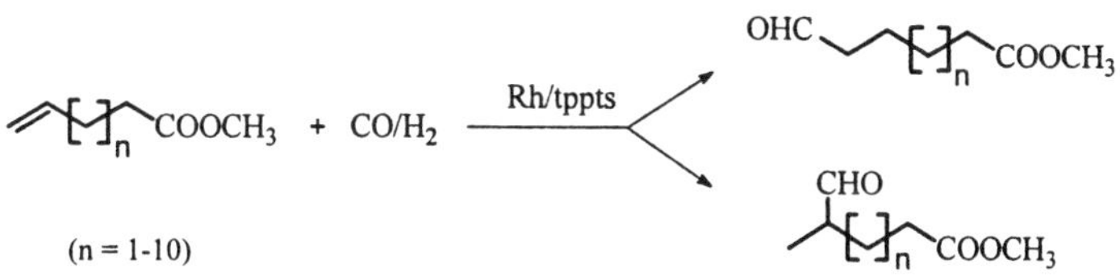

Figure 7 *Rh/tppts-catalysed hydroformylation of terminally unsaturated methyl esters.*

The polarity of the lower and mid range terminally unsaturated esters up to 9-decene carboxylic acid methyl ester (Figure 7, n≤6) provided for a Rh/tppts catalysed biphasic hydroformylation with acceptable rates. In contrast, 10-undecene- and 13-tetradecene-carboxylic acid methyl esters (Figure 7, n>6) reacted slowly but reaction rates could be increased by adding cationic surfactants such as tetradecyltrimethylammonium bromide.[89,417] Anionic and non-ionic surfactants were less effective. To rationalise the positive effect of cationic tensides it was assumed that the concentration of the negatively charged Rh/$P(C_6H_4\text{-}m\text{-}SO_3^-)_3$ catalyst is increased in the polarity gradient between the *Stern layer* and the *core* of the micelle, where the olefin is solubilised (cf. Figure 3) probably due to beneficial electronic interactions with the positively charged surface of the micelle.[89,389,417] Fell *et al.*[89] discussed the question of the form of micelles in which hydroformylation takes place and favored the "normal" micelle rationale (see Section 3.4.1 and Figures 3, 4). The presence and the type of the tenside also influenced the selectivity. For example, the n/i ratios obtained in the hydroformylation of 10-undecenenoate to the linear, 11-formylundecanoate and branched 10-formylundecanoate were 92/8 in the absence of tensides, 86/14 with anionic, 80/20 with zwitterionic and 70/30 with cationic surfactants.[89,417] The catalyst was recovered by phase separation and rhodium carry-over was below the detection limits (<1ppm).[417] Micellar Rh/tppts catalysed two phase hydroformylation was also used to convert polyunsaturated fatty esters such as the methyl esters of linolenic acid [$CH_3(CH_2CH{=}CH)_3(CH_2)_7CO_2Me$], linoleic acid [$CH_3(CH_2)_4CH{=}CH\text{-}CH_2\text{-}CH{=}CH\text{-}(CH_2)_7CO_2Me$], or oleic acid [$CH_3(CH_2)_7\text{-}CH{=}CH\text{-}(CH_2)_7CO_2Me$] as well as fatty oils like linseed oil to the corresponding formyl derivatives which are of considerable industrial interest.[89,419] Selectivities up to 53% to the triformyl derivative of methyl linolenate (most desired product), 44% to the diformyl derivative of methyl linoleate and 40% to monoformyl derivative of methyl oleate with quantitative conversions were achieved in the micellar Rh/tppts catalysed biphasic hydroformylation of technical grade methyl linolenate (55% linolenic, 16% linoleic and 20% oleic acid methyl esters) in the presence of cationic tensides such as $Me(CH_2)_{13}N^+Me_3\ Br^-$.[89] After the reaction the catalyst was easily separated by decantation and after washing of the organic phase the recovery was 99.5-99.9%. A coupled adsorption of the catalyst on silica gel reduces rhodium losses to the level of ppb, which is satisfactory from an industrial point of view.[89,419] The hydroformylation of unsaturated fatty esters was extensively investigated by the US Department of Agriculture in the period 1969-1984 using conventional rhodium and cobalt catalysts modified with triphenylphosphine or triphenyl phosphite ligands.[89,420,421] No commercial process emerged from this work due to difficulties in the separation and recycling of the catalyst in an active form as well as in the quantitative recovery of the rhodium.[89,421]

DSM jointly with Du Pont de Nemours[308] have patented platinum catalysts generated from the water soluble sulfonated ligand **30** (Table 2; m=0, n=0, m=1, n=2; m=1, n=1, Ar=^{n}Bu-SO_3Li) and used in the aqueous phase hydroformylation of internally unsaturated carboxylic acids, esters or nitriles to their corresponding formyl derivatives which are useful intermediates for the preparation of dicarboxylic acids (e.g. adipic acid). For example, TOFs up to 105 h^{-1} were achieved in the hydroformylation of 3-pentenoic acid catalysed by Pt/**30** (m=0, n=0) at 100°C and 80 bar CO/H_2 to give aldehydes with a selectivity of 83% (n/i=3.4), valeric acid (4.6%) and adipic acid (8.1%).[308] The products were separated from the aqueous catalyst solution by extraction with ether. Five recycles of the aqueous catalyst solution showed that the Pt/**30** (m=0, n=0) catalyst retains its activity.

4 Olefin Hydrocarboxylation

Metal-catalysed hydrocarboxylation of olefins (Equation 3) is the poor relative of the more famous hydroformylation. It generally requires forcing reaction conditions and suffers from mediocre activities and selectivities (n/i ratio). Moreover, the same products can be made via hydroformylation and oxidation of the aldehyde product.[431] Consequently, there are few industrial applications of hydrocarboxylation e.g. $Ni(CO)_4$-catalysed production of propionic acid by hydrocarboxylation of ethylene.[432,433]

$$\text{R-CH=CH}_2 + \text{CO} + \text{H}_2\text{O} \rightarrow \text{R-CH}_2\text{-CH}_2\text{-CO}_2\text{H} + \text{R-CH(CH}_3\text{)-CO}_2\text{H} \quad (3)$$

Palladium complexes such as $PdCl_2(PPh_3)_2$ catalyse hydrocarboxylations under milder conditions than nickel or cobalt catalysts.[431,434-445] Recently, $Pd(tppts)_3$ has been shown[452] to catalyse the hydrocarboxylation of olefins, in biphasic, aqueous/organic media under relatively mild conditions, in the presence of a *p*-toluenesulfonic acid cocatalyst. For example, surprisingly high activities (TOF>2500 h^{-1}) were observed in the hydrocarboxylation of propene (propene/Pd=1000; tppts/Pd=4, p-$CH_3C_6H_4SO_3H$/propene=0.148, 120°C and 50 bar CO) to give a 60/40 mixture of n-butyric and isobutyric acid. The activity of $Pd(tppts)_3$ was strongly dependent on the nature of the added Brønsted acid. Acids of weakly coordinating anions, e.g. p-$CH_3C_6H_4SO_3H$ or CF_3CO_2H, were effective while virtually no reaction was observed with acids of strongly coordinating anions, e.g. HI. The activity observed with $Pd(tppts)_3$ in a biphasic medium is substantially higher than those observed with Pd/PPh_3 in organic solvents.[439-442] (e.g. TOF=220 h^{-1} in aqueous dioxane).[439] This exceptionally high activity exhibited by Pd/tppts contrasts with the general perception that biphasic catalysis exhibits lower rates compared to analogous reactions in organic media.

Similarily, styrene and 4-isobutylstyrene were converted to the corresponding carboxylic acids (Equations 4 and 5) by reaction with CO and H_2O in the presence of $Pd(tppts)_3$ and p-$CH_3C_6H_4SO_3H$ (no organic solvent) at 65°C and 50-140 bar CO (the relatively low temperature is required to suppress polymerization of the substrate). The major products were the branched acids in both cases: 90% and 74% with styrene and 4-isobutylstyrene, respectively. In the case of 4-isobutylstyrene the branched acid product is the anti-inflammatory drug ibuprofen.

The activities observed in the hydrocarboxylation of styrene and 4-isobutyl-

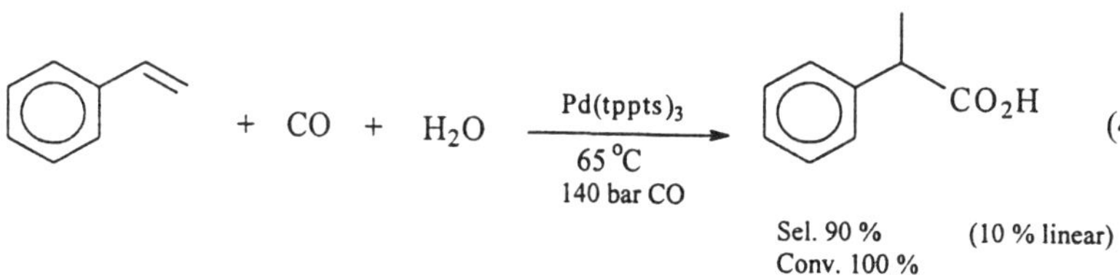

(4)

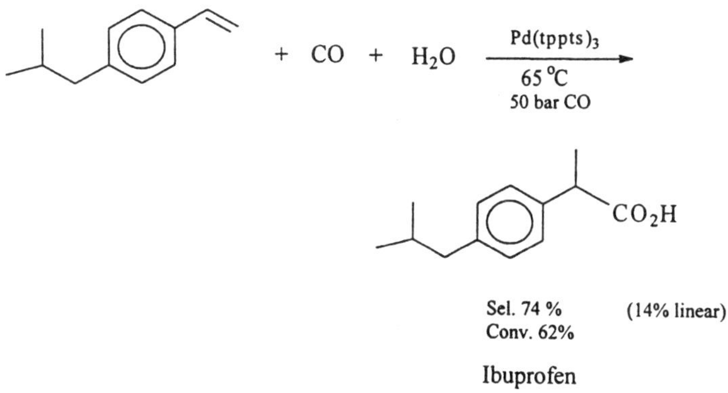

(5)

styrene (TOF= 49 and 3 h^{-1}, respectively) were much lower than with propene (see above), presumably due to the lower temperature and the lower solubilities of the olefin substrates in water.

Importantly, the $Pd(tppts)_3$ catalyst, generated *in situ* by complexation of $PdCl_2$ with tppts in H_2O followed by reduction of the resulting $[PdCl(tppts)_3]^+$ complex with CO,[448-451] was stable under hydrocarboxylation conditions.[452] In contrast, the Pd/PPh_3 system in organic solvents is not completely stable with respect to palladium metal formation.

The catalytic cycle depicted in Figure 8 was proposed to explain the observed results. Oxidative addition of HX to L_3Pd^o forms the cationic palladium tppts hydride $[L_3Pd^{II}H]^+$ which is the catalytically active species.[453] Dissociation of a tppts ligand is followed, successively, by olefin coordination, olefin insertion into the Pd-H bond to give intermediate I, CO coordination giving II, migratory insertion and nucleophilic attack of H_2O on the acyl intermediate III to give carboxylic acid and, after coordination of a tppts ligand, regenerate the initial hydride intermediate. *Markovnikov* addition of Pd-H species to the olefin leads to the branched carboxylic acid and *anti Markovnikov* addition to the linear product. When X^- is a strongly coordinating anion, such as I^-, the free coordination site in I is occupied, inhibiting coordination of CO.

A possible explanation for the higher catalytic activity of Pd/tppts compared with Pd/PPh_3 is that the lower basicity of tppts facilitates nucleophilic attack of H_2O on the acyl intermediate III (Figure 8). Nucleophilic attack of ROH on organic soluble acyl palladium intermediates has been proposed to be the rate-limiting step in Pd-catalysed carbonylations.[454] An alternative explanation is that the catalytically active *cis*-form of the tppts cationic Pd intermediate I is stabilized by intramolecular association of sulfonate groups of neighboring tppts ligands via hydrogen bonding in water.[345]

The single-phase hydrocarboxylation of 1-heptene catalysed by palladium in combination with tppts, tppms or PPh_3 in aqueous dioxane or acetone has also been reported.[455] However, activities were low (TOF=1-17 h^{-1}) and, in the case of PPh_3, palladium black formation was observed.[455]

5 Carbonylation

5.1 Carbonylation of Alcohols – $Pd(tppts)_3$ catalyses the carbonylation of benzylic alcohols to the corresponding phenylacetic acids, in the presence of a Brønsted acid cocatalyst such as H_2SO_4 or p-$CH_3C_6H_4SO_3H$ in biphasic aqueous/organic media (no organic solvent).[305,451] For example, benzyl alcohol was converted to phenylacetic acid (Equation 6) and 1-(4-isobutylphenyl)ethanol (IBPE) to ibuprofen (Figure 9).

Pd/PPh_3 complexes catalyse the carbonylation of activated benzyl alcohols such as 4-hydroxybenzyl alcohol whereas they are inactive in the carbonylation of benzyl alcohol.[463] In sharp contrast, the water soluble $Pd(tppts)_3$ complex catalyses the carbonylation of benzyl alcohol to phenylacetic acid, equation (6), under mild reaction conditions employing an aqueous/organic two phase

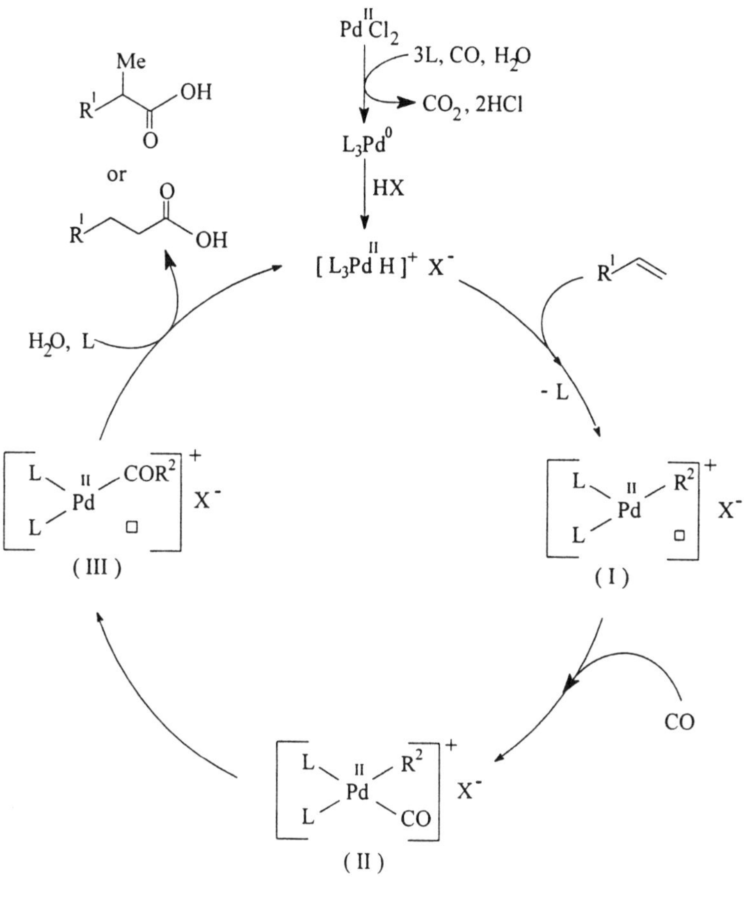

Figure 8 *Proposed catalytic cycle in the Pd(tppts)$_3$-catalysed hydrocarboxylation of olefins.*

system.[305] The best results (77% yield) were obtained at 100°C, 60 bar CO, tppts/Pd=12, and a palladium concentration of 150 ppm in the aqueous solution.

$$PhCH_2OH + CO \rightarrow PhCH_2COOH \qquad (6)$$

In the BHC (Boots-Hoechst Celanese) process about 3500 tons of ibuprofen per annum are produced by Pd/PPh_3-catalysed carbonylation of IBPE (Figure 9) in the presence of HCl, in organic media.[446,447,459-461] However, a shortcoming of this process is the cumbersome separation of the Pd/PPh_3 catalyst from the reaction products and its quantitative recovery in an active form as well as the highly corrosive reaction conditions due to the use of 10-26% aqueous HCl at reaction temperatures as high as 130°C. In the manufacture of pharmaceuticals the quantitative separation of the catalyst is even more important in order to avoid contamination of the product by trace amounts of heavy metals.

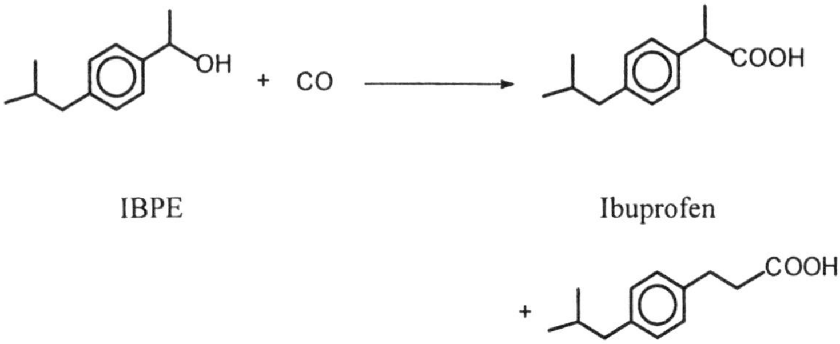

IBPE

Ibuprofen

3-IPPA

Figure 9 *Carbonylation of IBPE.*

$Pd(tppts)_3$-catalysed carbonylation of IBPE (Figure 9) in an aqueous/organic two phase system provides for facile separation of the catalyst from reaction products as well as its quantitative recovery and recycling in an active form.[451,462] Conversions up to 93% and selectivities to ibuprofen of 82% [18% linear product 3-IPPA (Figure 9)] were obtained at 90°C, 150 bar CO, and a palladium concentration of 150 ppm in the aqueous solution with no decomposition of $Pd(tppts)_3$ after the reaction.[451] Under these conditions the dehydration product 4-isobutylstyrene (IBS) was obtained in trace amounts. Both the activity and selectivity of carbonylation were strongly influenced by the tppts/Pd molar ratio and the nature of added protonic acid. Maximum efficiency was observed for tppts/Pd=10. High ibuprofen selectivities were obtained using acids of weaky coordinating anions such as CF_3COOH or p-$CH_3C_6H_4SO_3H$. The use of such non-corrosive acids is a further advantage of this biphasic reaction.[451] However, a disadvantage of the $Pd(tppts)_3$-catalysed IBPE carbonylation is the rather low catalytic activity (TOF=2.3 h^{-1}) compared to conventional Pd/PPh_3 system (TOF=950 h^{-1}).[451] The low catalytic activity may be due to the low water solubility of 4-isobutylstyrene (IBS). The latter, formed by acid-catalysed dehydration of IBPE, is probably an intermediate in the carbonylation reaction (see Section 4). In an independent experiment IBS was carbonylated eight times faster than IBPE. Moreover, the reaction temperature of 90°C is too low for facile dehydration of IBPE to IBS but is probably necessary for the stability of the carbonyl Pd/tppts intermediates involved in the catalytic cycle.

Replacement of tppts by a ligand containing less $-SO_3Na$ groups such as tppms gave rise to a dramatic drop in the catalytic activity and selectivity to ibuprofen. A palladium catalyst generated from the sulfonated diphosphine **29** (Table 2; x=3; m=n=0:86% and m=0,n=1:14%) exhibited low catalytic activity and the major product was 3-IPPA (78%).[451]

Similarly, 5-hydroxymethylfurfural (HMF) was chemoselectively carbonylated at the quasi-benzylic alcohol functionality to afford the previously undescribed 5-formylfuran-2-acetic acid (FFA) (Figure 10) as the sole carbonylation product, using the $Pd(tppts)_3$ catalyst[448-450] in a completely aqueous medium under mild reaction conditions (70°C, 5 bar CO).[305,449,458] The only byproduct observed was 5-methylfurfural (MF) resulting from a new type of catalytic and selective reduction with CO, formally equivalent to hydrogenolysis of an alcohol group without the presence of H_2 (Figure 10). HMF is of interest as a renewable resource derived from acid-catalysed conversion of carbohydrates such as fructose.[456,457] It is a potentially important basic raw material which can, in principle, be converted to a variety of interesting products.[457] For example, FFA is an interesting precursor for the manufacture of 2,5-furandiacetic acid and 5-carboxyfuran-2-acetic acid, which could form polymers e.g. polyesters and polyamides, the latter being comparable with the well-known terephthalic acid- and isophthalic acid-based polyamides.

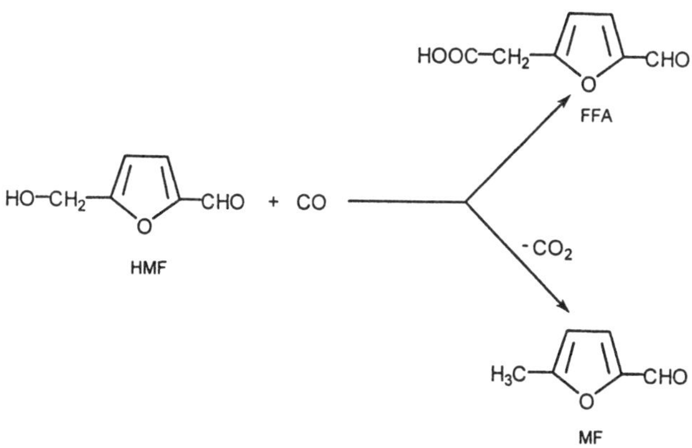

Figure 10 *$Pd(tppts)_3$-catalysed carbonylation and reduction of HMF.*

The carbonylation of HMF is the first example of a catalytic conversion of water soluble starting materials obtained from renewable resources using water soluble transition metal tppts complexes in aqueous media. Both the activity and selectivity of HMF carbonylation were strongly influenced by the tppts/Pd ratio; maximum efficiency being observed for tppts/Pd=6 (conversion=90%). The nature of the added acid markedly influenced the selectivity. Acids of weakly or non coordinating anions, such as CF_3COOH, p-$CH_3C_6H_4SO_3H$, H_2SO_4 and HPF_6 afforded mainly carbonylation (selectivity of FFA up to 83%). The

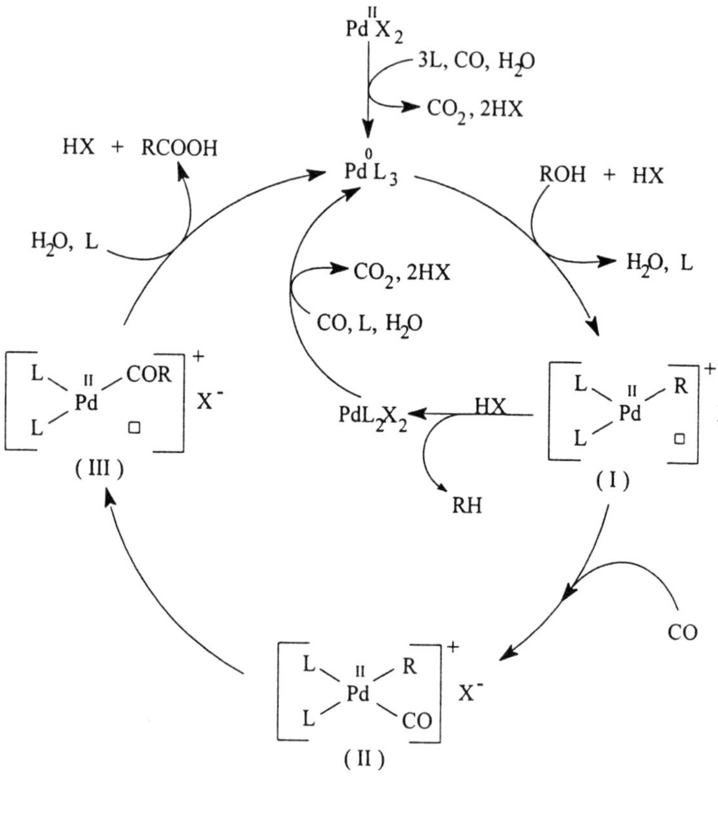

Figure 11 *Proposed catalytic cycle for the Pd(tppts)₃-catalysed carbonylation and reduction of HMF.*

selectivity decreased dramatically with acids of strongly coordinating anions such as HBr and HI. With the latter the only product observed was MF (selectivity=99.8%).[305,449] This high selectivity is even more remarkable when one considers that HMF contains a reactive aldehyde group.

In order to explain the observed results the catalytic cycle depicted in Figure 11 was proposed.[305,449] The $Pd(tppts)_3$ catalyst was generated *in situ* by reduction with CO of $[PdCl(tppts)_3]^+$ obtained via complexation of $PdCl_2$ in aqueous tppts solution.[448,450] This is followed, successively, by oxidative addition of RX to $Pd(tppts)_3$, CO coordination, migration of R to the carbonyl group of intermediate II with formation of the cationic acyl species III, and nucleophilic attack of H_2O on the carbon atom of III to give FFA and, at the same time regenerate the catalyst. When X^- is a strongly coordinating anion, coordination of CO is less favorable and intermediate (I) undergoes protonolysis to give RH and $Pd(tppts)_2X_2$. The latter undergoes subsequent reduction by CO to regenerate the palladium(0) catalyst.

Replacement of tppts by ligands containing less $-SO_3Na$ groups such as tppds or the disulfonated ligand **4** (Table 2, n=1) gives rise to a dramatic drop in the catalytic activity. Using palladium catalysts modified with tppms only traces of FFA and MF were obtained.

5.2 Carbonylation of Halides – $Pd(tppts)_3$-catalysed carbonylation of bromobenzene (Equation 7) in the presence of NEt_3 in an aqueous/toluene (1/1) two phase system at 150°C and 15 bar CO afforded the triethylammonium salt of benzoic acid (100% yield).[464,465] Rates were rather low (TOF's of 3.3-17 h^{-1}) but no decomposition of $Pd(tppts)_3$ (tppts/Pd $\geqslant$ 12.5) was observed and the catalyst could be recovered quantitatively and recycled.[464] However, in a second recycle extensive decomposition of the catalyst occurred with formation of palladium black. Generally in carbonylation reactions of halides the formation of stoichiometric amounts of either HX or halide salts still remains a problem of environmental concern despite the attractiveness due to the presence of the aqueous solvent.

$$PhBr + CO + H_2O + 2NEt_3 \rightarrow PhCOO^-N^+HEt_3 + NHEt_3Br \qquad (7)$$

Attempted reductive carbonylation of bromobenzene with CO/H_2 or with CO/HCOONa in the presence of a Pd/tppts catalyst failed to afford any benzaldehyde.[464]

Substituted benzyl chlorides were carbonylated using a Pd/tppts catalyst in aqueous/organic two phase systems under basic reaction conditions to afford the sodium salts of the corresponding phenylacetic acids. After acidification the phenylacetic acid dissolved in the organic phase and could be readily separated from the Pd/tppts catalyst contained in the aqueous phase (Figure 12).[466-468] TOFs up to 21 h^{-1} (turnover number, TON=165) and phenylacetic acid yields up to 94% were obtained at 70°C, 1 bar CO, tppts/Pd=10, NaOH/substrate=3/2 in an aqueous/toluene (1/1) two phase system in a batchwise procedure.[466] The TOFs were improved to a maximum of 135 h^{-1} (TON=1560) in a continuous operation mode. Palladium catalysts modified with binas (Table 2; **25**) exhibited low catalytic activity (TONs up to 140) in the carbonylation of benzyl chloride.[466] In strongly acidic media (pH=1) the Pd/**25** catalyst was active and remained stable during the reaction which contrasts with Pd/tppts where palladium black was observed. However, the catalyst was completely deactivated after three cycles.[466]

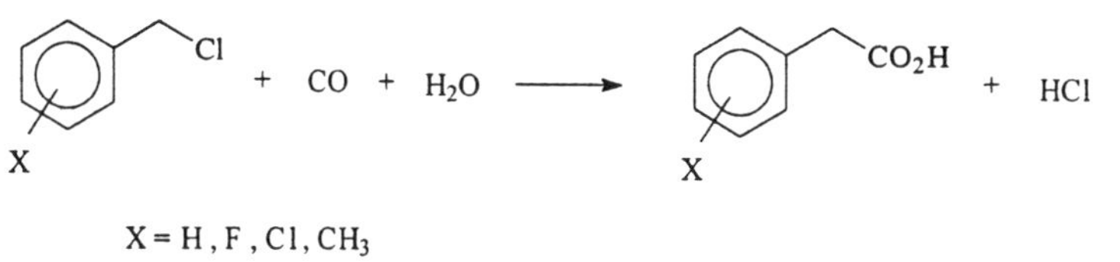

Figure 12 *Carbonylation of substituted benzyl chlorides.*

Benzyl chloride was carbonylated using $PdCl_2(tppms)_2$ in an aqueous/organic two phase system under basic conditions to give phenylacetic acid (Figure 12; X=H) in up to 93% yield, after a short induction period, under mild reaction conditions (50°C, atmospheric pressure of CO).[55,469] Addition of amphiphilic reagents considerably reduced the induction period and accelerated the carbonylation reaction.[55,470] For example, in the presence of n-$C_7H_{15}SO_3Na$ and n-$C_7H_{15}CO_2Na$ the absorption of CO was 90% and 87%, respectively compared to 14% in the absence of amphiphilic reagents.[55,470]

Similarly, $PdCl_2$ modified with the water soluble phosphonated phosphine **105** (Table 5) was used as catalyst in the carbonylation of benzyl chloride to afford phenylacetic acid in 91% yield in an aqueous/organic two phase system under mild reaction conditions (55°C, 8 psi CO).[259]

$PdCl_2(tppts)_2$ exhibited high catalytic activity (TOF=1860 h^{-1}) for the carbonylation of allyl chloride to the desired product β,γ-butenoic acid [Equation (8)] at 40°C and atmospheric CO pressure in an aqueous(NaOH)/toluene two phase system[471] By-products were α,β-butenoic acid and allyl alcohol. The yield of butenoic acids was 94% with a $\beta,\gamma/\alpha,\beta$ ratio of 21.[471]

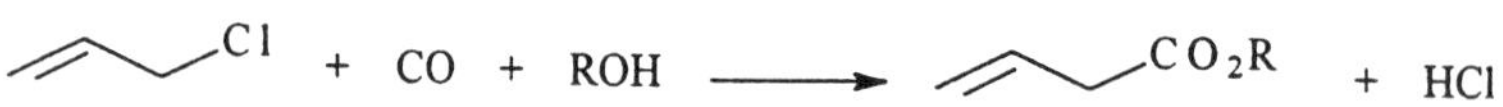

(8)

(R = H, Me, Et)

A synergistic effect on activity and selectivity was observed with polymer-anchored palladium based bimetallic catalysts containing tppts ligands in the carbonylation of allyl chloride to β,γ-butenoic acid at 40°C and atmospheric CO pressure in an aqueous/toluene two phase system.[472] For example, using $PdCl_2$ anchored on poly(N-vinyl-2-pyrrolidone) (PVP) with an aqueous solution of $RhH(CO)(tppts)_3$ the yield to butenoic acids was 95% and the $\beta,\gamma/\alpha,\beta$ ratio 13.3 compared to 48% yield and a $\beta,\gamma/\alpha,\beta$ ratio of only 4.9 with PVP-$PdCl_2$ alone.[472] Such synergistic effects were not observed using other transition metal compounds such as $Co(OAc)_2 \cdot 4H_2O$ or $NiCl_2 \cdot 6H_2O$, etc as a second component in the aqueous solution.[472]

$PdCl_2(tppms)_2$ was also used as a catalyst for the atmospheric pressure carbonylation of allyl halides at 30-50°C in the presence of sodium hydroxide or alkoxides to afford β,γ-unsaturated acids or esters in yields up to 93% ($\beta,\gamma/\alpha,\beta$ ratio 87/13) in biphasic or monophasic methanolic media.[55,56,473,474]

Carbonylation of phenethyl bromide catalysed by $Co_2(CO)_8$ alone or modified either with tppts or tppms yielded products of double- and mono-carbonylation, (benzylpyruvic and benzylacetic acid, respectively) at a concentration of 8 mol% cobalt, 85°C and 20 bar CO under basic conditions in an H_2O/tBuOH (14/1) solvent mixture (Figure 13).[475,476] The reaction rates exhibited by $Co_2(CO)_8$ were comparable with the $Co_2(CO)_6(tppts)_2$ catalyst but replacement of tppts by tppms gave rise to a dramatic decrease in catalytic activity.[475] Substantially different double/mono carbonylation (d/m ratios) were observed with $Co_2(CO)_8$, $Co_2(CO)_6(tppts)_2$ and $Co_2(CO)_6(tppms)_2$: 7.3, 2.1 and 0.3, respectively, which

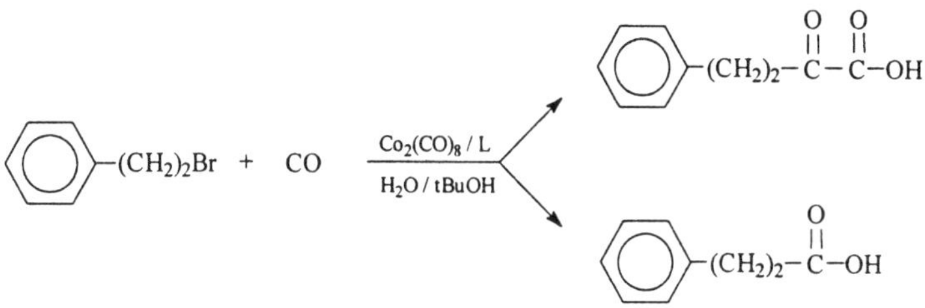

Figure 13 *Double- and mono-carbonylation of phenethyl bromide.*

were rationalised on the basis of increasing basicity of the ligands in the order: tppms>tppts>CO.[475]

6 Hydrogenation

The development of aqueous phase hydrogenation reactions catalysed by water soluble transition metal sulfonated phosphine complexes began with the pioneering work of Joó and Beck[480] in the Ru/tppms-catalysed hydrogenation of pyruvic acid to lactic acid in 1974 which coincides with the first application of water soluble $RhH(CO)(tppts)_3$ catalysts in the biphasic propene hydroformylation.[30] The success of Ru/tppms contrasted with the lack of activity observed using transition metal complexes modified with water soluble ligands such as the system **57** [Table 4, n=1] in hydrogenation reactions in aqueous media.[184,481]

6.1 Hydrogenation of Carbonyl Compounds – Rhône-Poulenc[54,485-487] has developed a new process for the biphasic Ru/tppts-catalysed hydrogenation of α,β-unsaturated aldehydes (important intermediates for vitamins, flavor and fragrance chemicals) to the corresponding allylic alcohols (Figure 14, II) in extremely high selectivities at quantitative conversions under mild reaction conditions. Hydrogenation of 3-methyl-2-buten-1-al to 3-methyl-2-buten-1-ol, for example, proceeds with 100% conversion and 97% selectivity at 35°C, 20 bar H_2 pressure and 1 hour reaction time[54,486] in an aqueous/organic (1/1) system. The catalyst was recycled twice without loss of selectivity. The addition of buffers in order to maintain neutral pH conditions inhibited the hydrogenation of the olefinic double bond which is favored in an acidic medium. The catalyst was prepared *in situ* from $RuCl_3$ and tppts in H_2O and after the reaction was easily separated by decantation. Studies of the catalyst showed that a stoichiometric redox reaction takes place between Ru(III), tppts and H_2O leading to the reduction of Ru(III) and oxidation of tppts.[485,486]

Ir/tppts catalysts exhibit almost the same selectivity as Ru/tppts in the hydrogenation of α,β-unsaturated aldehydes albeit with approximately 70 times lower rates.[485] In sharp contrast to the ruthenium and iridium based tppts catalysts, Rh/tppts complexes catalyse the chemoselective hydrogenation of α,β-unsaturated aldehydes to the corresponding saturated aldehydes (Figure 14, III).[54,485]

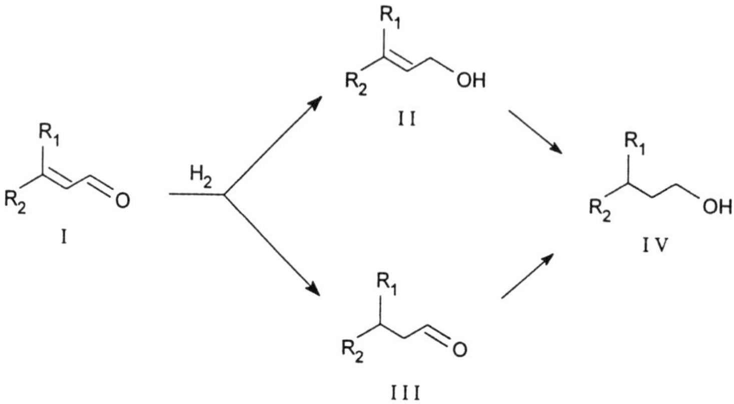

Figure 14 *Biphasic hydrogenation of α, β-unsaturated aldehydes.*

Hydrogenation of all-*trans*-retinal to all-*trans*-retinol[54,489] proceeds with high stereo- and chemoselectivity (97%) at quantitative conversions (TOF>338 h^{-1}) using a $H_2Ru(PPh_3)_4$ catalyst in a mixture of CH_3OH/H_2O=50/1 as solvent (Figure 15). In this case water soluble $H_2Ru(tppts)_4$ catalysts gave poor results in a two phase system presumably due to the insolubility of this substrate in water.

Ru/tppms catalysts exhibited excellent yields (98%) in the transfer hydrogenation of unsaturated aldehydes such as cinnamaldehyde or crotonaldehyde, to the corresponding unsaturated alcohols under mild reaction conditions (30-80°C), with HCOONa in an aqueous/organic two phase system.[490,491] Similarly, ruthenium modified with the water soluble, air stable phosphine **100** (pta; Table 5) is an effective catalyst for the chemoselective transfer hydrogenation of α,β-unsaturated aldehydes to unsaturated alcohols using formate in an aqueous/organic two phase system.[245,247,248,261] In contrast, Rh/pta afforded the corresponding saturated aldehydes.[247,248] The catalysts could be recycled in the aqueous phase with minimal leaching of rhodium.[248]

Ru/tppms and Ru/pta are also active catalysts for the hydrogenation of a series of substituted benzaldehydes to the corresponding benzyl alcohols with sodium formate as the source of hydrogen.[247,261,491]

Hydrogenation of *trans*-cinnamaldehyde (Figure 14; I; R_1=H, R_2=Ph) to phenylpropanol (IV; R_1=H, R_2=Ph) catalysed by Ru/tppms and Ru/tppts proceeds via two different routes in an aqueous/organic two phase system.[492] With $HRuCl(CO)(tppms)_3$ the reaction proceeds exclusively via the hydrocinnamaldehyde III whereas with $HRuCl(CO)(tppts)_3$ both II and III are involved (Figure 14, R_1=H, R_2=Ph).[492]

Osmium complexes modified with tppts and tppms also catalyse the selective hydrogenation of cinnamaldehyde to cinnamylalcohol (Figure 14, II) in an aqueous/organic two phase system.[493] The selectivity towards the unsaturated alcohol II with Os/tppts was lower than with Ru/tppts but both Ru and Os/tppts

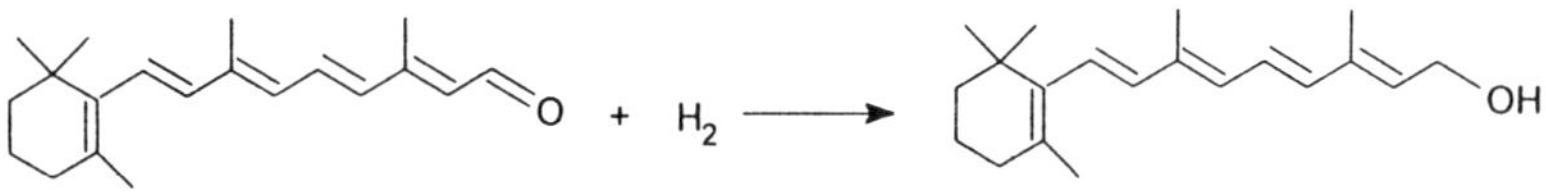

Figure 15 *Hydrogenation of all-trans-retinal to all-trans-retinol.*

catalysts showed higher selectivities than their PPh_3 modified homogeneous counterparts.[488,493]

Hernandez and Kalck[495,496] synthesized various Ru/tppts complexes and used them as catalysts in the hydrogenation of various α,β-unsaturated aldehydes and ketones in a two phase system. Using $Ru(H)_2(tppts)_4$ high conversions of cinnamaldehyde (97%) and high selectivity to cinnamylalcohol (95%) were observed under mild reaction conditions (40°C, 20 bar, 3h).[495] More forcing conditions (80°C, 35 bar, 16h) were needed for the hydrogenation of unsaturated α,β-ketones such as 4-hexen-3-one or benzylideneacetone and only low chemoselectivities were observed.[495]

Hydrogenation of saturated ketones such as cyclohexanone or 2-butanone catalysed by Ru/tppts complexes gave saturated alcohols in 100% selectivity at 5-99% conversion.

Basset *et al.*[482-484] showed that Ru/tppts catalysts exhibit extremely high catalytic activities in the selective aqueous phase hydrogenation of propionaldehyde to propanol under mild reaction conditions (35°C) and that the activity is dramatically enhanced in the presence of salts, especially of sodium iodide. Similarly, the hydrogenation of the epimeric aldoses, D-glucose and D-mannose obtained from renewable resources was recently reported using Ru/tppts catalysts at 100°C, 50 atm H_2 in the presence of sodium iodide (NaI/Ru=100) in aqueous media.[46,494] D-mannose was more reactive (TOF=800 h^{-1}) than D-glucose (TOF=520 h^{-1}).[494] In the absence of sodium iodide the rates obtained in the hydrogenation of D-mannose were six times lower. Ru/tppts catalysts were also active in the hydrogenation of D-mannose using NEt_3/HCOOH (2/5) as hydrogen source. However, the activity of this system is strongly pH dependent; maximum efficiency being observed at pH=8.0-8.4. An attempt to combine heptamolybdate-catalysed epimerization of D-glucose to D-mannose with subsequent Ru/tppts-NEt_3/HCOOH catalysed hydrogenation failed because the rate of epimerization decreased dramatically (ca. 20 times) by raising the pH only from 2.5 to 2.9 and at low pH Ru/tppts-NEt_3/HCOOH catalytic systems exhibited very low catalytic activity.[494]

6.2 Hydrogenation of Olefins

6.2.1 Hydrogenation of Simple Olefins – Dror and Manassen[497] hydrogenated cyclohexene using Rh/tppms catalysts at 50°C, tppms/Rh=6, in a dimethylacetamide/H_2O (2.5/1) solvent mixture. The catalyst was separated from the colourless cyclohexene phase indicating no leaching of rhodium. The authors[497] demonstrated that the reaction does not takes place at the interface by doubling of the volume of the aqueous/cosolvent phase which gave rise to a (small) increase in the reaction rate. If the hydrogenation were an interface reaction, the rates would not depend on the volume of the aqueous phase.[497] In contrast, Andriollo *et al.*[498] found a linear dependence of the reaction rate on the rate of stirring in the range 0-1000 rpm in the biphasic $RuH(CO)Cl(tppms)_3$-catalysed hydrogenation of cyclohexene (TOF=1.1 h^{-1}) and styrene (TOF=3.0 h^{-1}) suggesting reaction at the interface. We note, however, that in the Ru/tppts-catalysed biphasic hydrogenation of α,β-unsaturated aldehydes[485] a linear dependence of the reaction rate on the rate of stirring is observed only in the range of 500-1500 rpm. At higher stirring rates this dependence becomes nonlinear which militates against reaction at the interface.

Dror and Manassen[169] hydrogenated α- and cyclic olefins such as 1-octene, 1-dodecene and cyclohexene in micellar systems using rhodium catalysts modified with water-insoluble carboxylated tenside phosphines **45** (Table 3; n=3,5,7,9,11) in the presence of conventional tensides such as sodium dodecylsulfate (SDS) or cetyltrimethylammonium bromide (CTAB) and cosolvents e.g. dimethyl sulfoxide. Linear olefins were more reactive than cyclic olefins. Maximum efficiency was observed in the presence of the anionic surfactant SDS when **45** contained a chain of 5-7 carbon atoms. In contrast, using the cationic tenside CTAB the ligand **45** with 5 C atoms was almost inactive but became active again with dodecyl trimethylammonium bromide.

Rhodium catalysts generated from the sulfonated phosphine **23** (Table 2) were effective in the hydrogenation of olefins in an aqueous/organic or in a homogeneous methanol system, substantially higher rates being observed in the latter system.[131] For example, the TOF observed in the hydrogenation of 1-hexene in the biphasic system was 220 h^{-1} compared to 7860 h^{-1} in methanol.[131]

Rhodium catalysts modified with carboxylated phosphines **45** (Table 3; n=5, n=7)[229] and phosphonium phosphines **103** (Table 5; n=2,3,6,10)[255] form very active catalytic systems for the hydrogenation of olefins in aqueous/organic two phase systems.

6.2.2 Hydrogenation of Allyl Alcohol – Bergbreiter *et al.*[206] hydrogenated allyl alcohol using water soluble rhodium catalysts modified with the "smart ligand" **69** (Table 4) in aqueous media at 0°C. On heating the sample to 40-50°C the hydrogenation stopped but on cooling to 0°C hydrogenation was resumed. This *ca.* 400 fold change in reaction rate is due to an inverse temperature dependency of the catalyst solubility. At higher temperatures (40-50°C) the ligand **69** loses its hydration shell, becomes water insoluble and the reaction stops. Lowering of the temperature (to 0°C) causes the hydration shell to be reversibly restored, the Rh/**69** system redissolves in water and hydrogenation of the water soluble substrate is resumed (cf. Section 3.4.2. Thermoregulated Phase Transfer Catalysis).[206]

6.2.3 Enantioselective Hydrogenation of Prochiral Olefins – The presence of SDS increased both the rate and the enantioselectivity of hydrogenation of prochiral dehydroaminoacid derivatives using rhodium catalysts modified with the diphosphine **75a** (Table 4) in aqueous media.[500] For example, addition of 0.13 mmol of SDS to the Rh/**75a** catalyst shortened the reaction half time ($t_{1/2}$) from

390 min to 6 min with concomitant increase of the enantioselectivity from 83% e.e. (S) to 97% e.e. (S).[500] The increase in activity and enantioselectivity could be attributed to micellar effects since these changes occurred when the concentration of the SDS exceeded the CMC.[500]

Hanson *et al.*[149] hydrogenated the prochiral olefin methyl α-acetamidocinnamate using rhodium catalysts modified with the tenside chiral sulfonated diphosphine **34** (Table 2) in an ethylacetate/H_2O micellar system at 25°C and 1 bar H_2. The yield (100%) and enantiomeric excess (69%) were considerably higher than with the tetrasulfonated diphosphine **31** (Table 2; m=0, n=0) which gave 32% yield and 20% e.e. and the reaction time was shorter (1.5 versus 20h). Rh/**34** and Rh/**31** (m=0, n=0) gave nearly the same results (100% yield and 72-75% e.e. within $\leq$ 1h) in homogeneous methanol solutions.[149]

Wan and Davis[135,138] modified rhodium complexes with the water soluble chiral tetrasulfonated binap ligand **26** (Table 2) and used them as catalysts in the asymmetric hydrogenation of 2-acetamidoacrylic acid in aqueous media. The e.e. observed in neat water using Rh/**26** was approximately the same as that obtained with the unsulfonated Rh/binap in ethanol (68-70% versus 67%).[135]

6.3 Hydrogenation of CO_2 – In recent years carbon dioxide has attracted attention as a raw material as an ecologically and economically attractive alternative carbon source in chemicals manufacture. Gassner and Leitner[502,503] hydrogenated CO_2 to formic acid (Equation 9) using $RhCl(tppts)_3$ and achieved very high turnover numbers (TON=3440) at room temperature, 40 atm H_2/CO_2 (1/1) within 12h in an $HNMe_2$ containing aqueous solution.

$$H_2 + CO_2 \rightarrow HCO_2H \qquad (9)$$

This is a rare example of efficient organometallic catalysis in aqueous media which exhibits higher rates than conventional transition metal complexes in organic solvents. Other such examples are the Pd/tppts catalysed hydrocarboxylation of propene (cf. Section 4) and the (SAP) Rh/tppts catalysed hydroformylation of methyl acrylate (cf. Sections 3.5 and 11).

7 Alkylation

7.1 Addition of Active C-H Compounds to Dienes: the Rhône-Poulenc Process for Geranylacetone - Geranylacetone is a precursor of isophytol, a key intermediate in the manufacture of vitamine E (tocopherol) (see Figure 16), the world market of which is 10 000 t/a at a price of 25-30 $/kg tocopherylacetate.[54]

Rhône-Poulenc has commercialized an attractive new process for the manufacture of geranylacetone in which the key step is Rh/tppts-catalysed addition of methyl acetoacetate to the conjugated 1,3-diene system of myrcene in an aqueous/organic two phase system in the presence of methanol (Figure 17).[54,505] Excellent regioselectivity (>99%) is observed, at high myrcene conversion (98%), to give the β-ketoesters A/B in a ratio of 45/55.[54,504] Rh/tppts catalysts are less effective in aqueous/organic two phase media without addition of cosolvents such as

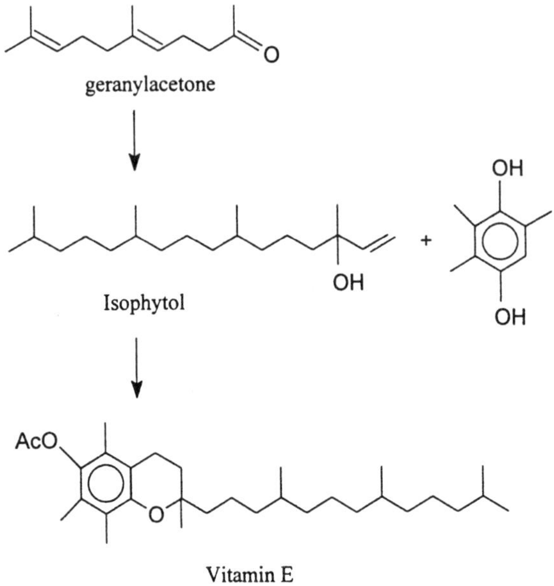

Figure 16 *Synthesis of vitamin E.*

methanol.[505] Patin *et al.*[505] replaced the sodium cation of tppts by NEt_4^+ or NBu_4^+ and generated rhodium catalysts which exhibited comparable rates in neat aqueous media with those achieved in H_2O/MeOH solvent mixtures.[505] The Rh/tppts-NBu_4^+ system in neat water was easily recycled. On the other hand, replacement of the sodium cation of the tetrasulfonated diphosphine **29** (Table 2; x=4, m=0, n=0) by NEt_4^+ or NBu_4^+ did not yield more effective catalysts and total loss of activity was observed on recycling.[505]

The raw material, myrcene (15 FF/Kg), is significantly cheaper than linalool (40 FF/Kg), the feedstock in the existing process.[54] Furthermore, the intermediate β-ketoesters can be used for the synthesis of pseudoionone a key intermediate for the manufacture of vitamin A (world market for vitamin A: 3 000 t/a at a price of 40-50 $/kg).[54,504]

The Rh/tppts catalysed C-C coupling reaction (Figure 17) has a broad scope and numerous conjugated dienes e.g. isoprene and β-farnesene can be used as starting materials with other active methylene compounds e.g. acetylacetone and barbituric acid.[506] Figure 18 depicts the Rh/tppts/Na_2CO_3-catalysed isoprenylation of barbituric acid which proceeds at room temperature in an H_2O/MeOH (75/25) solvent mixture.[506]

7.2 Heck-, Suzuki- and Stille-type Couplings – The ***Heck reaction***, palladium-catalysed coupling of aryl or vinyl halides with alkenes or alkynes, is an extremely useful synthetic method. Only recently have Heck reactions been performed in aqueous media, probably due to the perception that water must be carefully

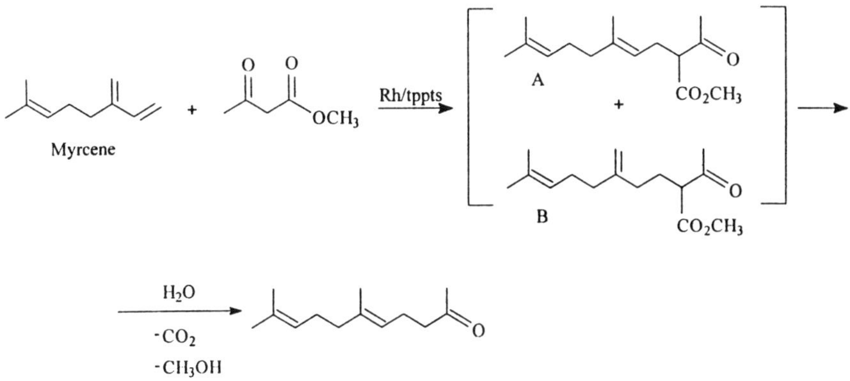

Figure 17 *The novel biphasic Rh/tppts-catalysed Rhône-Poulenc process for the manufacture of geranylacetone.*

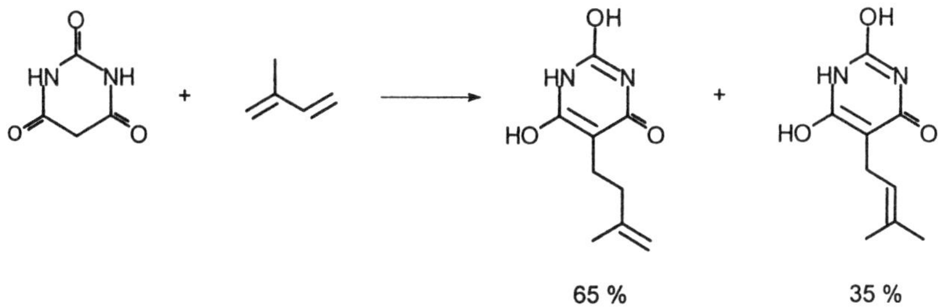

Figure 18 *Isoprenylation of barbituric acid.*

avoided in order to exclude palladium-catalysed Wacker-type oxidations of alkenes to aldehydes and/or ketones.[55,507,508] Both inter-[44,55,56,233,507,509,510] and intra-molecular[511-513] Heck reactions in aqueous media have been reported. For example, Kiji *et al.*[55,56,507] used $PdCl_2(tppms)_2$ to catalyse the coupling of various aryl halides with ethylene to give styrenes (Figure 19) at 100°C and 20-50 atm ethylene in the presence of NEt_3 or $NaHCO_3$ in H_2O/THF or H_2O/CH_3CN solvent mixtures.

Both aqueous $NaHCO_3$-soluble and-insoluble aryl bromides bearing electron-withdrawing groups afforded styrenes in good to quantitative yields, whereas aryl bromides with electron-donating groups reacted slowly. Aryl iodides were, as expected, more reactive than bromides.[507] The $PdCl_2(tppms)_2$ catalyst was easily recovered by simple phase separation.[55] Palladium black was formed when the catalyst was added separately as $Pd(OAc)_2$ and tppms instead of as preformed $PdCl_2(tppms)_2$ complex.[507]

Similarly, $Pd(OAc)_2$ modified with the guanidino phosphines **92** (Table 5; n=1,2) catalysed the Heck reaction between 4-iodobenzoic acid and 4-carboxy-

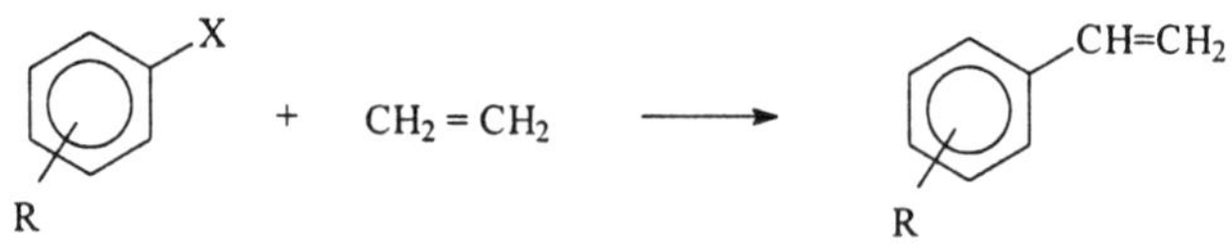

X = I, Br
R = CO_2H, CO_2Me, OMe, CN, NO_2, H, Cl, Br

Figure 19 *Heck arylation of ethylene in aqueous media.*

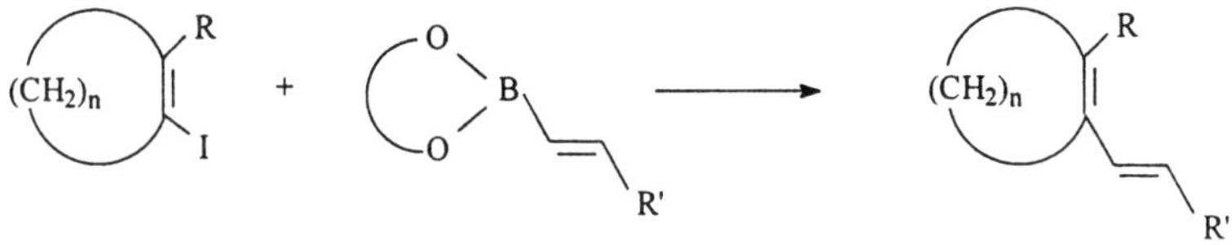

R = CO_2Me
R' = Bu, $CH(OEt)_2$, $CH(OAc)CH_3$

Figure 20 *Suzuki-type cross coupling reaction.*

phenyl acetylene to afford cross- and homo-coupling products in the presence of CuI in H_2O or H_2O/CH_3CN solvent mixtures albeit with a lower activity than Pd/tppts.[233] However, we note that cationic guanidinium phosphines **92** (n=1,2) are more stable towards air oxygen than sulfonated arylphosphines.[233]

Suzuki-type cross coupling reactions of vinyl iodides with alkenylboron reagents (Figure 20) were also studied in the system containing Pd/tppts at room temperature in the presence of $HN(iPr)_2$ in H_2O/CH_3CN solvent mixtures.[514] Interestingly, this coupling reaction proceeds with retention of Z olefin geometry and formation of functionalised dienes and trienes.[514]

$PdCl_2$-, $PdCl_2(tppms)_2$- or Pd/tppds-catalysed ***Stille-Couplings*** of water soluble aryl and vinyl halides with alkyl-, aryl- or vinyltrichlorostannates (Figure 21) proceed in aqueous alkaline media to give cross-coupled products at 90-100°C with good to excellent yields.[515,516]

Pd/tppts-catalysed intermolecular sp-sp couplings to afford diynes and sp^2-sp coupling of 2-iodophenols or iodoanilines with alkynes followed by intramolecular cyclization to benzofurans and indoles, respectively, with moderate to good yields in H_2O/CH_3CN solvent mixtures have also been described.[517]

8 Telomerization of 1,3-Butadiene: the Kuraray Process

Telomerization is one of the very first reactions carried out in aqueous/organic two phase systems. In 1976, Kuntz at Rhône-Poulenc[518] reduced $PdCl_2$ with $NaBH_4$ in the presence of tppms to afford a catalyst for the telomerization of

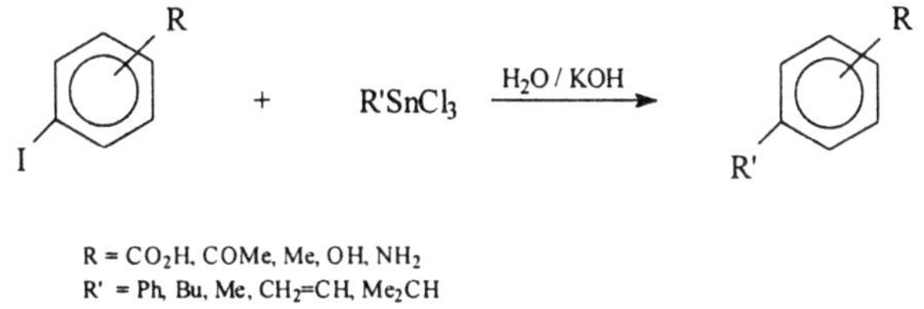

Figure 21 *Stille-type coupling reaction.*

butadiene with methanol, at 30°C in the presence of KOH, to give 95% 1-methoxy-2,7-octadiene and 4% of 3-methoxy-1,7-octadiene. Similarily, $Pd(OAc)_2$/tppts-catalysed telomerization of 1,3-butadiene in the presence of aqueous K_2CO_3 at 80°C afforded 63% of 2,7-octadiene-1-ol and 21% of 1,7-octadiene-3-ol.[518]

Kuraray[59,74,425-430,572] has commercialized a new process for the production of 1,9-nonanediol (2-3000 tpa)[59] via Pd/tppms-catalysed telomerization of 1,3-butadiene with water to 2,7-octadien-1-ol (Figure 22, I) followed by isomerization (II) to 1-octen-7-al over a $CuCrO_3$ catalyst, Rh/tppms-catalysed hydroformylation (III) and nickel-catalysed hydrogenation (IV) of the resulting 1,9-nonanedial.[427] The Kuraray process[59,74,425-430,572] is also used for producing 1-octanol (5000 tpa)[59] via nickel-catalysed hydrogenation of 2,7-octadiene-1-ol (Figure 22, V).[428,572] The telomerization of 1,3-butadiene is catalysed by palladium complexes generated, for example, from tppms or alkenylphosphonium derivatives of Li-tppms at 75°C in the presence of amines such as NEt_3 under an atmosphere of CO_2 in sulfolane/water mixtures[428,429] to give 2,7-octadiene-1-ol and 1,7-octadiene-3-ol in a ratio of 167/9.[74,430,572] The hydroformylation of 7-octenal is carried out with Rh/tppms at 110°C and 90 atm CO/H_2 to give a conversion of 97% within 6 h.[426] The Rh/tppms catalyst is separated from the 1,9-nonanedial product by decantation after the addition of water and recycled after evaporation of the aqueous phase.[426] Alternatively, the hydroformylation reaction can be carried out in the presence of polyethylene glycol dimethylethers (mol wt. $\geq$300) under milder reaction conditions e.g. 90°C within 4 h.[425] The Rh/tppms catalyst is separated from the 1,9-nonanedial by decantation after the addition of water and standing for 15h. The Rh/tppms catalyst with the entire amount or a part of the polyethylene glycol derivative is recycled after evaporation of the aqueous phase.[425] 1,9-Nonanedial is a useful intermediate not only for the production of 1,9-nonanediol but also to produce nonanedioic acid (azelaic acid) and nonane diamines.[425,426,572]

Mortreux *et al.*[519] telomerized 1,3-butadiene with water to 2,7-octadiene-1-ol (selectivity up to 94%) using Pd/tppts catalysts in the presence of amines and CO_2 (10 bar) at 85°C in an aqueous/organic two phase system without any addition of organic solvents. TOFs up to 269 h^{-1} were achieved only when the added amine contains a long alkyl chain, e.g. $C_{12}H_{25}NMe_2$. The authors proposed that the amine reacts with carbonic acid (from CO_2 and H_2O) to give ammonium cationic surfactants capable of forming micelles in which the telomerization reaction takes place.[519] Interestingly, catalytic activities of TOF=245 h^{-1} were obtained with

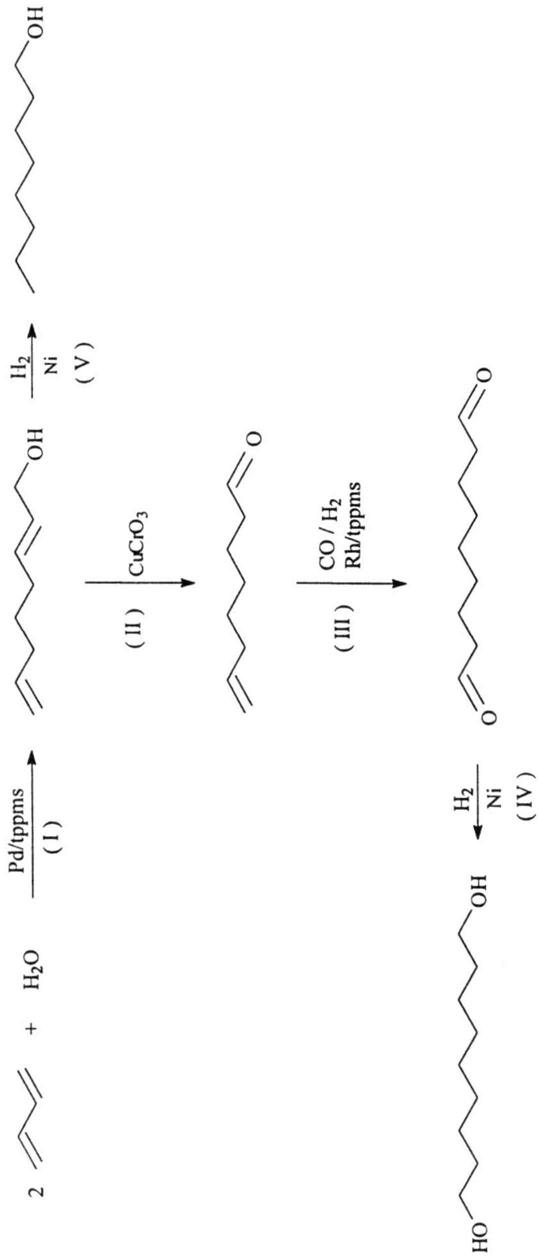

Figure 22 *The Kuraray process for the production of 1,9-nonanediol or 1-octanol*

long chain tetraalkylammonium salts of weakly coordinating anions e.g. $C_{12}H_{25}NMe_3^+$ OH^- whereas the reaction was totally inhibited with ammonium salts of strongly coordinating anions such as chlorides or bromides.[519]

Recently, interesting effects were revealed in the Pd/tppts-catalysed telomerization of 1,3-butadiene and ammonia in an aqueous/organic two phase system (Figure 23).[522] The polarity of the aqueous phase in the biphasic Pd/tppts-catalysed reaction prevented a further reaction of the desired primary octadienylamines (Figure 23, I and II) to secondary (III) or tertiary (IV) products whereas in organic monophasic Pd/PPh_3-catalysed systems high yields of tertiary octadienylamines (IV) were obtained. For example, selectivities up to 88% to primary amines I and II were achieved in the Pd/tppts catalysed reaction in an aqueous/toluene two phase system while selectivities of *ca.* 60% to tertiary octadienylamines IV were common in the Pd/PPh_3 catalysed single phase tBuOH system.[522] The total yields obtained in the mono and two phase system were, however, moderate (21-61%). The catalyst was recycled five times without any decrease in selectivity albeit with a small decrease in activity. The novel concept of using water soluble catalysts to avoid consecutive reactions in aqueous/organic two phase systems may find broad application in the future in other fields e.g. hydroformylation of alkynes.

Palladium complexes generated from the sulfonated ligand **5** (Table 2; R=Me, n=0,1,2) were also used to catalyse the telomerization of 1,3-butadiene with water[108] and from the aminophosphines **83** (Table 5; R=Ph; R′=H,Me; R″=Me,iPr; n=2,3) with MeOH to give 1-methoxy-2,7-octadiene in selectivities up to 78% at conversions ranging from 52 to 80%.[232] Pd/**83** (R=Ph; R′=H,Me; R″=Me,iPr; n=2,3) catalysts are more active but less selective in the latter reaction compared to their tppts counterparts.[232]

9 Miscellaneous Reactions

Other examples of the application of organometallic catalysis in water are: ***Oligomerization*** of ethylene using Ni/tppms[523] and Ni/**20** (Table 2; n=0, R=Me,

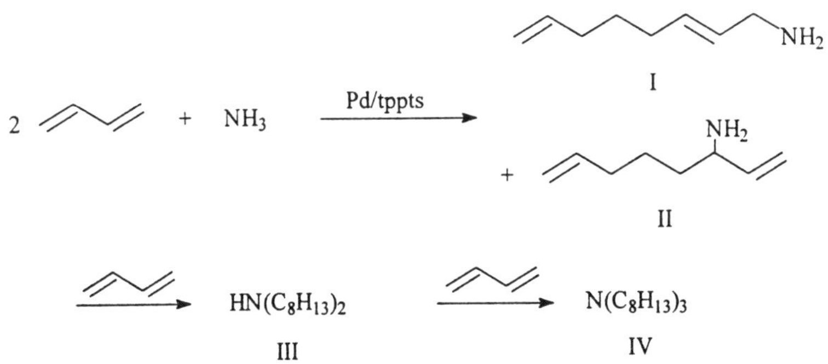

Figure 23 *Telomerization of 1,3-butadiene with ammonia.*

Na-salt; n=0, R=H, Li-salt) catalysts[524] (the latter was used in polar solvents such as ethanol and sulfolane) or of terminal alkynes catalysed by Rh/tppms and Rh/**42** (Table 3; n=2) systems.[525]

Ruthenium-catalysed ***Ring-Opening Metathesis Polymerization (ROMP)***[129,526-529] (Equation 10) and Pd(tppms)$_3$-catalysed synthesis of water soluble poly-(*p*-phenylene) derivatives (Equation 11).[530] The latter is a special example of a Suzuki coupling (see earlier).

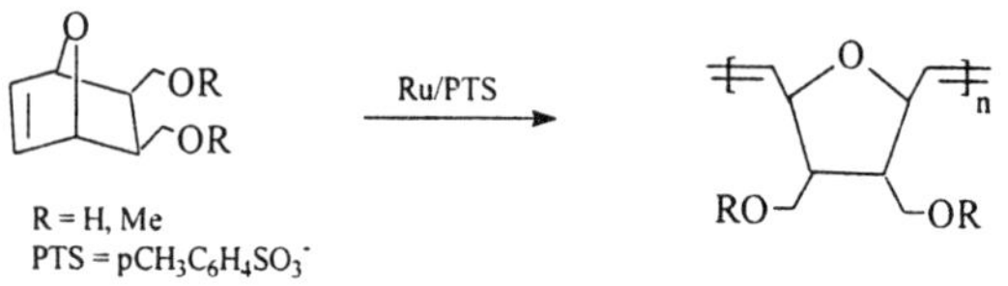

(10)

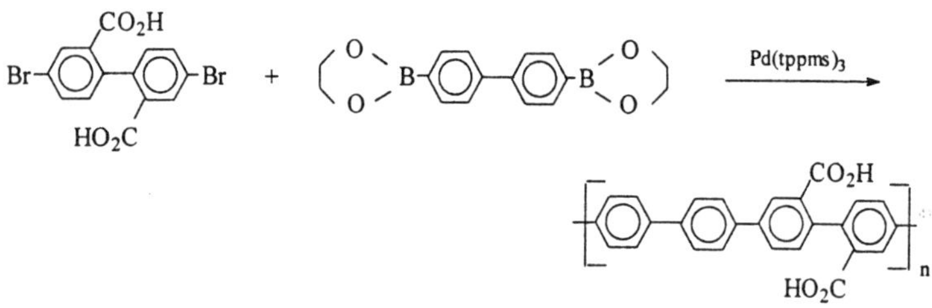

(11)

Alternating Copolymerization of Olefins with CO catalysed by palladium complexes generated from the sulfonated ligands **29** (Table 2; x=3; m=0,1; n=0) and **108** (Table 6) in aqueous media[263] or from **20** (Table 2; n=0; R=H; acid-form) and **21** (Table 2; R=Ph; n=2,3; acid-form) in methanol.[335]

Allylic Substitutions with a variety of nucleophiles catalysed by Pd/tppts[147,510,531-533] or Pd/**29** (Table 2; x=2,4; m=0; n=0)[531,532] in aqueous-organic media (Equation 12).

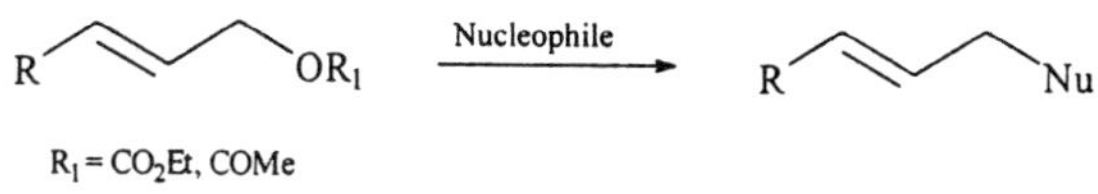

(12)

Pd/tppts-catalysed ***allylations*** of aromatic amines (Equation 13) and phenols[49,534] or uracils and 2-thiouracils.[535]

(13)

Selective ***removal of allyl and allyloxycarbonyl protecting groups*** promoted by Pd/tppts catalytic systems (Equation 14).[43,536-540]

Hydrocyanation of 1,3-butadiene and 3-pentene nitrile to adiponitrile (Equation 15) catalysed by Ni/tppts systems in the presence of reducing agents, such as $NaBH_4$, or $ZnCl_2$.[49,541,542]

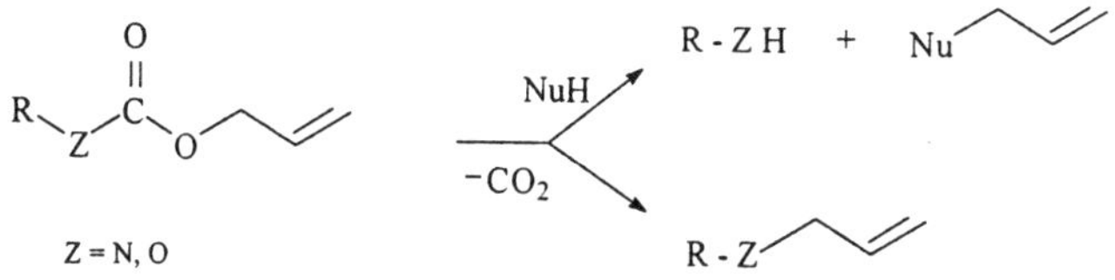

(14)

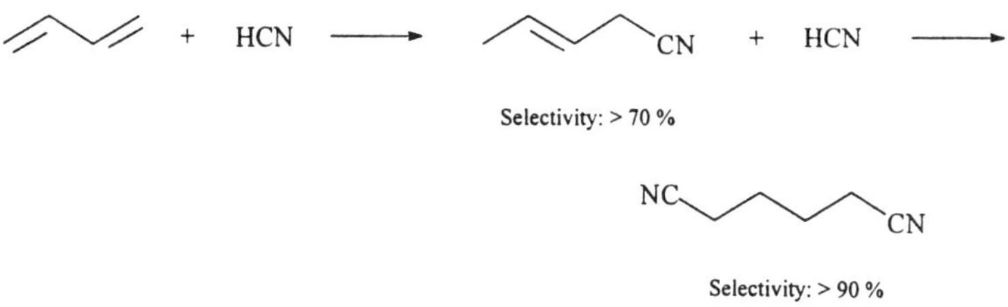

(15)

Oxidations catalysed by Co/**122** (Table 6)[317-319], Mn/**126** (Table 6; R=Me, CH_2CH_2OH)[332] and $PdCl_3(pyridine)^-$ systems[546,551] or ruthenium complexes[543] and $CoBr_2$[544] in micellar systems. We note also that the classical Wacker process employing $PdCl_2/CuCl_2$ catalytic systems constitutes another application of oxidation reactions in aqueous media.[552]

Reduction of nitro aromatics to anilines with CO (Equation 16) catalysed by palladium complexes generated from tppts or **25** (Table 2)[47,547,548,550b] or by $[Rh(CO)_4]^-$ systems[549,550a] in aqueous media.

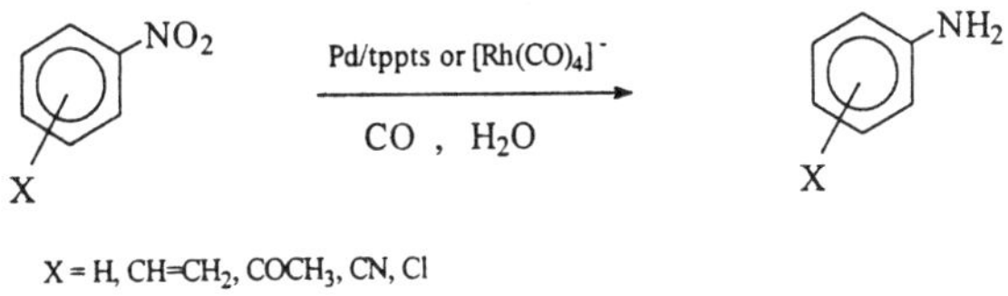

(16)

Isomerization of allylbenzene and 1-octene-3-ol catalysed by various metals (Rh, Ru, Ir) modified with tppts, tppms and the carboxylated phosphines, depicted in Table 3, **41** (n=2), **42** (n=0,1,2) and **43** as well as with **100** (Table 5).[248,553,554]

Epimerization reactions of D-glucose and D-mannose catalysed by various transition metals (Cu, Ni, Co etc.) modified with the ligands **113** (Table 6).[272]

Hydrodehalogenation of organic halides catalysed by Ru/tppms[555] or palladium complexes generated from the ligands **20** (n=1, R=H) and **21** (R=Ph, n=3,4) (depicted in Table 2).[124,125]

Asymmetric hydrogenolysis of epoxides catalysed by Rh/**31** (Table 2; m=n=0; m=0, n=1; m=n=1; m=1, n=2) complexes.[556]

Hydrolysis of activated esters of α-amino acids or acetylcholine analogs catalysed by Cu or Zn modified with the tenside ligands **118**, **119** and **120** (Table 6).[276-279]

Finally, although not directly related to organometallic catalysis, we mention that ***water soluble phosphines*** such as tppms or tppts react with activated olefins[166,557,558], alkynes[559,560], aldehydes[192,194,561] and methyl iodide[562] in aqueous media ***to give phosphonium salts***.

10 Counter (Inverse) Phase Transfer Catalysis

The term *Counter Phase Transfer Catalysis* (CPTC) was coined by Okano[214,215] to describe biphasic reactions catalysed by water soluble transition metal complexes which involve transport of an organic-soluble reactant into the aqueous phase where the catalytic reaction takes place. Similarly, Mathias and Vaidya[564,565] gave the name *Inverse Phase Transfer Catalysis* to describe this kind of biphasic catalysis which contrasts with classical *Phase Transfer Catalysis* where the reaction occurs in the organic phase and does not involve formation of micelles.[389,564]

Classical Phase Transfer assisted Organometallic Catalysis[588-601] is a further important field which has found industrial applications e.g. in the carbonylation of benzyl chloride to phenylacetic acid using $NaCo(CO)_4/Bu_4NBr$ catalysts in aqueous NaOH practiced by Montedison.[52,466,589,596,601] However, a detailed discussion of classical phase transfer catalysis is beyond the scope of this chapter which is devoted to systems in which the catalytic conversion takes place in the aqueous phase.

The concept of CPTC has been applied in a large number of catalytic reactions such as reduction of allyl chlorides with HCOONa, carbonylation of aryl and allyl halides, allylation of aldehydes, cyanation of aryl halides etc.[214] For example, Okano *et al.*[215] reduced 1-chloro-2-nonene to afford 1-nonene and 2-nonene (Equation 17) with selectivities of 82% and 17%, respectively using palladium catalysts modified with the water soluble polyether phosphine **76** (Table 4; n=0, x=1, m=3, R=Me) in an heptane/H_2O (8/2) two phase system according to CPTC concept.

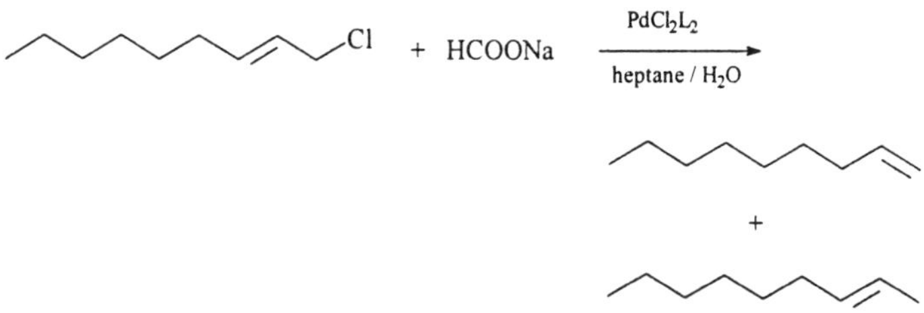

(17)

With $PdCl_2(\mathbf{76})_2$ the conversion of 1-chloro-2-nonene was quantitative whereas with the non-polar $PdCl_2[P(nBu)_3]_2$ the lipophilic substrate hardly came into contact with the hydrophilic reductant HCOONa and the conversion was low (26%).[215]

Rh/tppts catalysts in the presence of cyclodextrins constitute CPTC systems which are active in the hydroformylation of various water-insoluble olefins.[566,567,570] For example, 1-decene was biphasically hydroformylated with quantitative conversions and up to 95% yields to aldehydes (n/1=1.9) using Rh/tppts with chemically modified β-cyclodextrins at 80°C and 50 atm CO/H_2 pressure.[566,570] After the reaction the two phases were well separated. Leaching of Rh and P into the organic phase was < 0.5 and 1.2 ppm, respectively.[566]

We note, however, that addition of α-cyclodextrins decreased the yields in the hydroformylation of 1-hexene catalysed by $RhH(CO)(tppms)_3$ complexes in a two phase system.[568,569]

$PdSO_4/H_9PV_6Mo_6O_{40}/CuSO_4$ in the presence of chemically modified β-cyclodextrins were also used as CPTC systems in the Wacker-type oxidation of higher α-olefins (C_8-C_{16}) to the corresponding 2-ketones (Equation 18) with high yields (90-98%) in an aqueous/organic two phase system.[545,571]

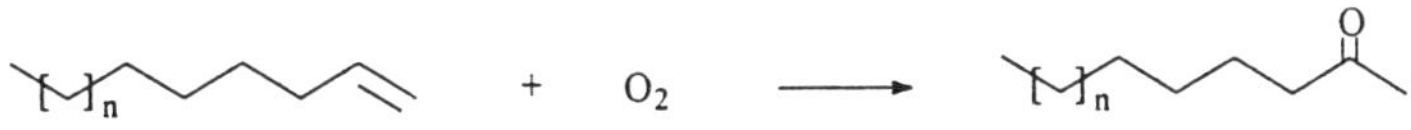

(18)

(n = 1-9)

11 Supported Aqueous Phase (SAP) Catalysis

An attractive approach to heterogenizing homogeneous catalysts is to immobilize water soluble organometallic complexes in a thin aqueous film on a high surface area hydrophilic support, such as silica or controlled pore glasses, to yield Supported Aqueous Phase (SAP) catalysts.[573-580] Surprisingly high TOFs (up to 4300 h^{-1}) with high yields (95%) to aldehydes (i/n=136) were obtained in the hydroformylation of methyl acrylate using SAP-Rh/tppts catalysts which were even 19 times more active than conventional Rh/PPh_3 complexes in toluene.[423,581] The content of water of the SiO_2 support played a key role in the activity of such SAP-Rh/tppts catalysts; maximum efficiency being observed with a water content of 37 wt.%. The catalytic activity dropped after recycling of the catalysts which was attributed to rhodium leaching and gradual dehydration of silica.[581] Replacement of the polar methyl acrylate substrate by the non-polar propene did not give such spectacular increase of reaction rates using SAP-Rh/tppts catalysts.[423,581] This suggests that the very high catalytic activity exhibited by such systems in the hydroformylation of methyl acrylate is not a general phenomenon. The authors[423] proposed that polar interactions between the polar methyl acrylate and the hydrated silica gel are responsible for the high catalytic activity.

A particular advantage of SAP catalysis[573-580] is that reaction with hydrophobic substrates occurs at the aqueous/organic interface allowing water insoluble higher olefins to be hydroformylated using the $RhH(CO)(tppts)_3$ system. In the biphasic hydroformylation using $RhH(CO)(tppts)_3$ the reaction rates decrease in the order: 1-hexene > 1-octene > 1-decene. In contrast, using the same catalyst in a silica supported aqueous phase[578] similar activities and selectivities were obtained in the hydroformylation of these olefins. The low n/i ratio observed in these reactions was explained by invoking the formation of $RhH(CO)_2$(tppts) species. Conversions up to 97% were achieved in the hydroformylation of oleyl alcohol using SAP-Rh/tppts catalysts at 100 °C, 51 bar CO/H_2 pressure and 5.5 hours reaction time[575] without significant leaching of rhodium into the organic phase.

SAP-Rh/tppts catalysts were used in the hydroformylation of methyl linolenate (technical grade) to give mainly mono- and di-formyl derivatives.[582] α,β-unsaturated aldehydes were selectively hydrogenated into allylic alcohols using SAP-Ru/tppts catalysts at a pressure of 100 bar H_2, 50°C and 3 hours reaction time[583]. However, using apolar solvents the activity of the catalyst decreased on recycling which was attributed to catalyst poisoning by adsorption of organic compounds. In polar solvents, such as methanol, leaching of the metal is the primary cause of catalyst deactivation.[583]

SAP catalysts have also been applied in the Wacker oxidation[584] of higher olefins where the separation of products from the catalyst is cumbersome. Palladium(II) and copper(II) salts immobilized on controlled pore glass CPG-240 in the presence of water catalysed the oxidation of 1-heptene to 2-heptanone in conversions up to 24%.[585] Significant isomerization to 2-heptene and 3-heptene (isomerization/oxidation=2/3) was also observed. However, an advantage of SAP-Wacker oxidation catalysts over classical systems is that the Cu(II) is confined to the support and therefore not corrosive whereas aqueous Cu^{2+} is very corrosive to steel.

Modified SAP catalysts have also been described in which the water was replaced with ethylene glycol and successfully applied to enantioselective catalysis. For example, water soluble chiral Ru/**26** (Table 2) complexes dissolved in a thin film of ethylene glycol adsorbed on a controlled pore glass catalysed the enantioselective (96% e.e.) hydrogenation of an (S)-naproxen precursor dissolved in a chloroform/cyclohexane (1/1) mixture (Equation 19).[136,137,586] No leaching of ruthenium (detection limit 32 ppb) into the organic phase was observed. Interestingly, a lower enantioselectivity (70% e.e.) was achieved when the chiral Ru/**26** catalyst was contained in a thin aqueous film instead of ethylene glycol.[136,137,586]

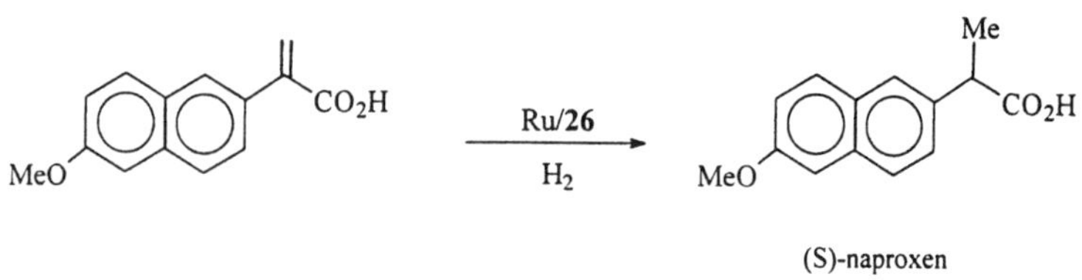

(19)

12 Concluding Remarks / Future Prospects

Organometallic catalysis in biphasic aqueous/organic media is a novel and environmentally benign approach to combining the inherent advantages of homogeneous catalysts (high activity and selectivity) with the ease of separation characteristic of heterogeneous catalysts. Several industrial applications, including the hydroformylation of lower olefins, the addition of active C-H compounds to dienes and the telomerization of 1,3-butadiene with water underscore the practical utility of this kind of heterogenization of homogeneous catalysis. The exceptionally high catalytic activities obtained in the Pd/tppts-catalysed hydrocarboxylation of propene, the (SAP)Rh/tppts-catalysed hydroformylation of methyl acrylate and the Rh/tppts-catalysed hydrogenation of CO_2 contrast with the general perception that catalysis by water soluble transition metal complexes normally exhibit lower rates compared to analogous reactions in organic media. A further advantage is improvement in selectivity by avoidance of consecutive reactions, e.g. in the telomerization of 1,3-butadiene with ammonia. Hence, we conclude that there is immense potential for further applications in the synthesis of fine chemicals. Catalytic conversions of highly water immiscible substrates using this technique still remains a challenge, however. Several conceptual advances such as micellar microheterogeneous systems, thermoregulated phase transfer catalysis and SAP catalysts are particularly elegant and potentially appealing approaches to circumventing this solubility problem. We note, however, that the practical utility of these various approaches stands or falls with the extent of leaching of the metal during numerous recycles. Finally, catalytic conversions of renewable carbohydrate resources is an emerging field which may benefit enormously from the application of aqueous phase organometallic catalysis.

References

1. R. A. Sheldon, *Curr. Opin. Solid State Mater. Sci.*, 1996, **1**, 101-106.
2. J. A. Moulijn, P. W. N. M. van Leeuwen, and R. A. van Santen "Catalysis: An Integrated Approach to Homogeneous, Heterogeneous and Industrial Catalysis", Elsevier, Amsterdam, 1993, p.v.
3. G. C. Pimentel, "Opportunities in Chemistry", NAS/NRC/02-86, National Academy Press, Washington, DC, 1985, p. 21.
4. G. C. Bond, "Heterogeneous Catalysis: Principles and Applications", Oxford University Press, Oxford, UK, 2nd edn, 1987, p. 8.
5. W. Keim, *Chem. Ing. Tech.*, 1984, **56**, 850-853.
6. A. Behr, *Fat Sci. Technol.*, 1990, **92**, 375-388.
7. C. N. Satterfield, "Mass Transfer in Heterogeneous Catalysis", M.I.T. Press, Cambridge, USA, 1970, p. 2.
8. J. C. Bailar, *Catal. Rev. Sci. Eng.*, 1974, **10**, 17-36.
9. W. Keim, *Stud. Surf. Sci. Catal.*, 1989, **44** (Success. Des. Catal.) 321.
10. W. Keim, T. M. Shryne, R. S. Bauer, H. Chung, P. W. Glockner, and H. van Zwet, Shell Int. Res., *DE 20 54 009*, 1969.
11. G. W. Parshall, "Homogeneous Catalysis: The Applications and Chemistry of Catalysis by Soluble Transition Metal Complexes", Wiley, New York, 1980, p.58-59.
12. I. T. Horváth and J. Rábai, *Science*, 1994, **266**, 72-75.
13. J. A. Gladysz, *Science*, 1994, **266**, 55-56.
14. I. T. Horváth and J. Rábai, Exxon Res. Eng. Co., *EP 633 062*, 1993; *Chem. Abstr.* 1995, **123**, 87349a; *US 5 463 082*, 1995.
15. Y. Chauvin, L. Mussmann, and H. Olivier, *Angew. Chem.*, 1995, **107**, 2941-2943; *Angew. Chem. Int. Ed. Engl.*, 1995, **34**, 2698-2700.

16. H. Olivier and Y. Chauvin, in "Catalysis of Organic Reactions", ed. R.E. Malz, Dekker, New York, 1996, p. 249-263.
17. Y. Chauvin and H. Olivier-Bourbigou, *Chemtech*, 1995, **25**, 26-30.
18. Y. Chauvin, S. Einloft, and H. Olivier, *Ind. Eng. Chem. Res.*, 1995, **34**, 1149-1155.
19. P. A. Z. Suarez, J. E. L. Dullius, S. Einloft, R. F. de Souza, and J. Dupont, *Polyhedron*, 1996, **15**, 1217-1219.
20. K. R. Seddon, in "Room-Temperature Ionic Liquids: Neoteric Solvents for Clean Catalysis", www: http://www.ch.qub.ac.uk/resources/ionic/review/review.html" (22.05.1997).
21. J. K. D. Surette, L. Green, and R. D. Singer, *Chem. Commun.*, 1996, 2753-2754.
22. G. W. Parshall, *J. Am. Chem. Soc.*, 1972, **94**, 8716-8719.
23. M. Poliakoff and S. Howdle, *Chem. Brit.*, 1995, **31**, 118-121.
24. T. Clifford and K. Bartle, *Chem. Ind. (London)*, 1996, 449-452.
25. M. Poliakoff, S. M. Howdle, and S. G. Kazarian, *Angew. Chem.* 1995, **107**, 1409-1432; *Angew. Chem. Int. Ed. Engl.*, 1995, **34**, 1275-1295.
26. J. W. Rathke, R. J. Klingler, and T. R. Krause, *Organometallics*, 1991, **10**, 1350-1355.
27. J. Manassen, in "Catalysis: Progress in Research", ed. F. Basolo and R. L. Burwell, Plenum Press, London, 1973, p. 177-185.
28. G. Papadogianakis, Ph.D. Thesis, Rheinisch-Westfälische Technische Hochschule Aachen, Germany, 1990, p. 35.
29. A. F. Borowski, D. J. Cole-Hamilton, and G. Wilkinson, *Nouv. J. Chim.*, 1978, **2**, 137-144.
30. E. G. Kuntz, *Chemtech*, 1987, **17**, 570-575.
31. B. Cornils and W. A. Herrmann, in "Applied Homogeneous Catalysis with Organometallic Compounds", ed. B. Cornils and W. A. Herrmann, VCH, Weinheim, 1996, vol. 2, ch.3, p. 575-601.
32. R. Gärtner, B. Cornils, H. Springer, and P. Lappe, Ruhrchemie AG, *DE 32 35 030*, 1982; *Chem. Abstr.* 1984, **101**, 55331t.
33. L. Bexten, B. Cornils, and D. Kupies, Ruhrchemie AG, *DE 34 31 643*, 1984; *Chem. Abstr.*, 1986, **105**, 117009n.
34. B. Cornils, J. Hibbel, W. Konkol, B. Lieder, J. Much, V. Schmidt, and E. Wiebus, Ruhrchemie AG, *EP 01 03 810*, 1983; *DE 32 34 701*, 1982; *Chem. Abstr.* 1984, **100**, 194022k.
35. G. Papadogianakis and R. A. Sheldon, *New J. Chem.*, 1996, **20**, 175-185.
36. B. Cornils and E. Wiebus, *Recl. Trav. Chim. Pays-Bas*, 1996, **115**, 211-215.
37. B. Cornils and E. Wiebus, *Chemtech*, 1995, **25**, 33-38.
38. E. Wiebus and B. Cornils, *Chem. Ing. Tech.*, 1994, **66**, 916-923.
39. R. A. Sheldon, *Chemtech*, 1994, **24**, 38-47.
40. R. A. Sheldon, in "Precision Process Technology", ed. M. P. C. Weijnen and A. A. H. Drinkenburg, Kluwer, Dordrecht, 1993, p. 125-138.
41. K. V. Katti, *Curr. Sci.*, 1996, **70**, 219-225.
42. E. Wiebus and B. Cornils, *Hydrocarbon Process. (Int. Ed.)*, 1996, **75**(3), 63-66.
43. S. Lemaire-Audoire, M. Savignac, J.-P. Genet and J.-M. Bernard, in "The Roots of Organic Development", ed. J.-R. Desmurs and S. Ratton, Industrial Chemistry Library, Elsevier, Amsterdam, 1996, vol. 8, p. 416-430.
44. J. Kiji, *Macromol. Symp.*, 1996, **105**, 167-171.
45. A. Lubineau, *Chem. Ind. (London)*, 1996, 123-126.
46. S. Kolari and V. Šunji, *J. Mol. Catal. A: Chem.*, 1996, **111**, 239-249.
47. A. M. Tafesh and J. Weiguny, *Chem. Rev.*, 1996, **96**, 2035-2052.
48. M. Beller, B. Cornils, C. D. Frohning, and C. W. Kohlpaintner, *J. Mol. Catal. A: Chem.*, 1995, **104**, 17-85.
49. B. Cornils and E. G. Kuntz, *J. Organomet. Chem.*, 1995, **502**, 177-186.
50. B. Cornils, *Angew. Chem.*, 1995, **107**, 1709-1711; *Angew. Chem. Int. Ed. Engl.*, 1995, **34**, 1575-1577.
51. D. M. Roundhill, *Adv. Organomet. Chem.*, 1995, **38**, 155-188.

52. R. V. Chaudhari, A. Bhattacharya, and B. M. Bhanage, *Catal. Today*, 1995, **24**, 123-133.
53. P. A. Chaloner, M. A. Esteruelas, F. Joó, and L. A. Oro, in "Homogeneous Hydrogenation", ed. R. Ugo and B. R. James, Catalysis by Metal Complexes, Kluwer, Dordrecht, 1994, vol. 15, ch. 5, p. 183-239.
54. C. Mercier and P. Chabardes, *Pure Appl. Chem.*, 1994, **66**, 1509-1518.
55. J. Kiji and T. Okano, *Rev. Heteroat. Chem.*, 1994, **11**, 191-206.
56. J. Kiji and T. Okano, *Yuki Gosei Kagaku Kyokaishi*, 1994, **52**, 276-284.
57. A. Chauvel, B. Delmon, and W. F. Hölderich, *Appl. Catal. A*, 1994, **115**, 173-217.
58. A. Lubineau, J. Augé, and Y. Queneau, *Synthesis*, 1994, 741-760.
59. N Nojiri and M. Misono, *Appl. Catal. A*, 1993, **93**, 103-122.
60. C.-J. Li, *Chem. Rev.*, 1993, **93**, 2023-2035.
61. W. A. Herrmann and C. W. Kohlpaintner, *Angew. Chem.*, 1993, **105**, 1588-1609; *Angew. Chem. Int. Ed. Engl.*, 1993, **32**, 1524-1544.
62. Ph. Kalck and F. Monteil, *Adv. Organomet. Chem.*, 1992, **34**, 219-284.
63. M. Barton and J. D. Atwood, *J. Coord. Chem.*, 1991, **24**, 43-67.
64. H. Papp and M. Baerns, in "New Trends in CO Activation", ed. L. Guczi, Studies in Surface Science and Catalysis, Elsevier, Amsterdam, 1991, vol. 64, ch. 10, p. 431-461.
65. T. G. Southern, *Polyhedron*, 1989, **8**, 407-413.
66. D. Sinou, *Bull. Soc. Chim. Fr.*, 1987, (3), 480-486.
67. J. P. Collmann, L. S. Hegedus, J. R. Norton, and R. G. Finke, "Principles and Applications of Organotransition Metal Chemistry", University Science Books, Mill Valley, California, 2nd edn, 1987, ch.12, p. 631-632.
68. F. Joó and Z. Tóth, *J. Mol. Catal.*, 1980, **8**, 369-383.
69. G. Süss-Fink, A. Meister, and G. Meister, *Coord. Chem. Rev.*, 1995, **143**, 97-111.
70. I. T. Horváth and F. Joó, in "Aqueous Organometallic Chemistry and Catalysis", ed. I. T. Horváth and F. Joó, NATO ASI Series, Series 3: High Technology, Kluwer, Dordrecht, 1995, vol. 5, p. 1-13.
71. J. Haggin, *Chem. Eng. News*, 1994, **72**(41), 28-36.
72. B. Cornils, *Nachr. Chem. Tech. Lab.*, 1994, **42**, 1136-1138.
73. I. T. Horváth, in Ref. 31, p. 601-605.
74. E. G. Kuntz, in Ref. 70, p. 177-181.
75. G. Papadogianakis, B. Fell, and H. Bahrmann, Hoechst AG, *EP 489 330*, 1991; *DE 9 016 585*, 1990; *Chem. Abstr.* 1992, **117**, 131385m.
76. B. Fell and G. Papadogianakis, *J. Prakt. Chem./Chem.-Ztg.*, 1994, **336**, 591-595.
77. T. Bartik, B. Bartik, B. E. Hanson, T. Glass, and W. Bebout, *Inorg. Chem.*, 1992, **31**, 2667- 2670.
78. H. Ding and B. E. Hanson, *J. Chem. Soc., Chem. Commun.*, 1994, 2747-2748.
79. H. Ding and B. E. Hanson, *J. Mol. Catal. A: Chem.*, 1995, **99**, 131-137.
80. T. Bartik, B. Bartik, B. E. Hanson, K. H. Whitmire, and I. Guo, *Inorg. Chem.*, 1993, **32**, 5833-5837.
81. H. Ding, B. E. Hanson, and T. E. Glasş, *Inorg. Chim. Acta*, 1995, **229**, 329-333.
82. W. A. Herrmann, G. A. Albanese, R. B. Manetsberger, P. Lappe, and H. Bahrmann, *Angew. Chem.*, 1995, **107**, 893; *Angew. Chem. Int. Ed. Engl.*, 1995, **34**, 811-813.
83. W. A. Herrmann, R. Manetsberger, G. Albanese, H. Bahrmann, P. Lappe, and K. Bergrath, Hoechst AG, *EP 632 047*, 1995; *Chem. Abstr.* 1995, **122**, 133420w.
84. G. Albanese, R. Manetsberger, and W. A. Herrmann, Hoechst AG, *EP 704 450*, 1996; *Chem. Abstr.* 1996, **125**, 11136z.
85. G. Albanese, R. Manetsberger, W. A. Herrmann, and Ch. Schwer, Hoechst AG, *EP 704 451*, 1996; *Chem. Abstr.* 1996, **125**, 11135y.
86. W. A. Herrmann, J. A. Kulpe, J. Kellner, H. Riepl, H. Bahrmann, and W. Konkol, *Angew. Chem.*, 1990, **102**, 408-414.
87. W. A. Herrmann, J. A. Kulpe, W. Konkol, and H. Bahrmann, *J. Organomet. Chem.*, 1990, **389**, 85-101.
88. W. A. Herrmann, J. Kellner, and H. Riepl, *J. Organomet. Chem.*, 1990, **389**, 103-128.
89. B. Fell, D. Leckel, and Ch. Schobben, *Fat Sci. Technol.*, 1995, **97**, 219-228.

90. C. Larpent, R. Dabard, and H. Patin, *Inorg. Chem.*, 1987, **26**, 2922-2924.
91. C. Larpent, R. Dabard, and H. Patin, *New J. Chem.*, 1988, **12**, 907-913.
92. C. Larpent and H. Patin, *Appl. Organomet. Chem.*, 1987, **1**, 529-534.
93. D. J. Darensbourg, Ch. J. Bischoff, and J. H. Reibenspies, *Inorg. Chem.*, 1991, **30**, 1144-1147.
94. D. J. Darensbourg and Ch. J. Bischoff, *Inorg. Chem.*, 1993, **32**, 47-53.
95. Ph. Kalck, P. Escaffre, F. Serein-Spirau, A. Thorez, B. Besson, Y. Colleuille, and R. Perron, *New J. Chem.*, 1988, **12**, 687-690.
96. P. Escaffre, A. Thorez, and Ph. Kalck, *New J. Chem.*, 1987, **11**, 601-604.
97. B. Fontal, J. Orlewski, C. C. Santini, and J.-M. Basset, *Inorg. Chem.*, 1986, **25**, 4320-4322.
98. S. Ahrland, J. Chatt, N. R. Davies, and A. A. Williams, *Nature (London)*, 1957, **179**, 1187- 1188.
99. S. Ahrland, J. Chatt, N. R. Davies, and A. A. Williams, *J. Chem. Soc.*, 1958, 276-288.
100. S. Ahrland, J. Chatt, N. R. Davies, and A. A. Williams, *J. Chem. Soc.*, 1958, 1403-1405.
101. G. Schmid, N. Klein, and L. Korste, *Polyhedron*, 1988, **7**, 605-608.
102. G. Schmid, *Polyhedron*, 1988, **7**, 2321-2329.
103. Y. Y. Yan, H.-P. Zuo, and Z.-L. Jin, *J. Nat. Gas Chem.*, 1996, **5**, 161-165.
104. I. Willner and R. Maidan, *J. Chem. Soc., Chem. Commun.*, 1988, 876-877.
105. M. J. Abrams, *Platinum Met. Rev.*, 1995, **39**, 14-18.
106. K. R. Seddon, *Platinum Met. Rev.*, 1996, **40**, 128-134.
107. G. Papadogianakis, in Ref. 28, p. 169-176.
108. E. Monflier, P. Bourdauducq, J.-L. Couturier, Elf Atochem S.A., *WO 95/30636*, 1995; *Chem. Abstr.* 1996, **124**, 201631j.
109. C. Larpent, H. Patin, N. Thilmont, and J. F. Valdor, *Synth. Commun.*, 1991, **21**, 495-503.
110. O. Herd, K. P. Langhans, O. Stelzer, N. Weferling, and W. S. Sheldrick, *Angew. Chem.*, 1993, **105**, 1097-1099; *Angew. Chem. Int. Ed. Engl.*, 1993, **32**, 1058-1059.
111. O. Herd, A. Hessler, K. P. Langhans, O. Stelzer, W. S. Sheldrick, and N. Weferling, *J. Organomet. Chem.*, 1994, **475**, 99-111.
112. O. Stelzer, K. P. Langhans, and N. Weferling, Hoechst AG, *DE 4141299*, 1991.
113. H. Schindlbaur, *Monatsh. Chem.*, 1965, **96**, 2051-2057.
114. F. Bitterer, O. Herd, A. Hessler, M. Kühnel, K. Rettig, O. Stelzer, W. S. Sheldrick, S. Nagel, and N. Rösch, *Inorg. Chem.*, 1996, **35**, 4103-4113.
115. B. Fell and G. Papadogianakis, *J. Mol. Catal.*, 1991, **66**, 143-154.
116. G. Papadogianakis in Ref. 28, p. 65-107.
117. H. Ding, B. E. Hanson, T. Bartik, and B. Bartik, *Organometallics*, 1994, **13**, 3761-3763.
118. B. E. Hanson, H. Ding, T. Bartik, and B. Bartik, in Ref. 70, p. 149-158.
119. T. Bartik, H. Ding, B. Bartik, and B. E. Hanson, *J. Mol. Catal. A: Chem.*, 1995, **98**, 117-122.
120. T. Bartik, B. Bartik, B. E. Hanson, I. Guo, and I. Tóth, *Organometallics*, 1993, **12**, 164-170.
121. T. Bartik, B. Bartik, and B. E. Hanson, *J. Mol. Catal.*, 1994, **88**, 43-56.
122. T. Bartik, B. Bartik, I. Guo, and B. E. Hanson, *J. Organomet. Chem.*, 1994, **480**, 15-21.
123. G. Oehme, in "Coordination Chemistry and Catalysis" ed. J. J. Ziolkowski, World Scientific, Singapore, 1988, p. 269-293.
124. E. Paetzold, G. Oehme, and B. Costisella, *Z. Chem.*, 1989, **29**, 447-448.
125. E. Paetzold and G. Oehme, *J. Prakt. Chem./Chem.-Ztg.*, 1993, **335**, 181-184.
126. E. Paetzold, A. Kinting, and G. Oehme, *J. Prakt. Chem.*, 1987, **329**, 725-731.
127. S. Kanagasabapathy, Z. Xia, G. Papadogianakis, and B. Fell, *J. Prakt. Chem./Chem.-Ztg.*, 1995, **337**, 446-450.
128. S. Ganguly, J. T. Mague, and D. M. Roundhill, *Inorg. Chem.*, 1992, **31**, 3500-3501.
129. B. Mohr, D. M. Lynn, and R. H. Grubbs, *Organometallics*, 1996, **15**, 4317-4325.

130. G. Fremy, Y. Castanet, R. Grzybek, E. Monflier, A. Mortreux, A. M. Trzeciak, and J. J. Ziolkowski, *J. Organomet. Chem.*, 1995, **505**, 11-16.
131. R. Grzybek, *React. Kinet. Catal. Lett.*, 1996, **58**, 315-322.
132. W. A. Herrmann, Ch. W. Kohlpaintner, R. B. Manetsberger, H. Bahrmann, and H. Kottmann, *J. Mol. Catal. A: Chem.*, 1995, **97**, 65-72.
133. W. A. Herrmann, G. Albanese, R. Manetsberger, R. Schmid, and C. Schwer, in Ref. 70, p. 127-129.
134. W. A. Herrmann, Ch. W. Kohlpaintner, H. Bahrmann, and W. Konkol, *J. Mol. Catal.*, 1992, **73**, 191-201.
135. K. T. Wan and M. E. Davis, *J. Chem. Soc., Chem. Commun.*, 1993, 1262-1264.
136. K. T. Wan and M. E. Davis, *Nature (London)*, 1994, **370**, 449-450.
137. K. T. Wan and M. E. Davis, *J. Catal.*, 1994, **148**, 1-8.
138. M. E. Davis and K. T. Wan, California Institute of Technology, *WO 95/22405*, 1995.
139. T. Bartik, B. B. Bunn, B. Bartik, and B. E. Hanson, *Inorg. Chem.*, 1994, **33**, 164-169.
140. L. Lecomte and D. Sinou, *Phosphorus, Sulfur, Silicon*, 1990, **53**, 239-251.
141. M. Laghmari and D. Sinou, *J. Mol. Catal.*, 1991, **66**, L15-L18.
142. C. Lensink and. J. G. de Vries, *Tetrahedron Asym.*, 1992, **3**, 235-238.
143. Y. Amrani, L. Lecomte, and D. Sinou, *Organometallics*, 1989, **8**, 542-547.
144. J. Bakos, in Ref. 70, p. 231-244.
145. F. Alario, Y. Amrani, Y. Colleuille, T. P. Dang, J. Jenck, D. Morel, and D. Sinou, *J. Chem. Soc., Chem. Commun.*, 1986, 202-203.
146. L. Lecomte and D. Sinou, *J. Mol. Catal.*, 1989, **52**, L21-L24.
147. D. Sinou, in Ref. 70, p. 215-230.
148. J. Bakos, A. Orosz, B. Heil, M. Laghmari, P. Lhoste, and D. Sinou, *J. Chem. Soc., Chem. Commun.*, 1991, 1684-1685.
149. H. Ding, B. E. Hanson, and J. Bakos, *Angew. Chem.*, 1995, **107**, 1728-1730; *Angew. Chem. Int. Ed. Engl.*, 1995, **34**, 1645-1647.
150. R. G. Nuzzo, D. Feitler, and G. M. Whitesides, *J. Am. Chem. Soc.*, 1979, **101**, 3683-3685.
151. R. G. Nuzzo, S. L. Haynie, M. E. Wilson, and G. M. Whitesides, *J. Org. Chem.*, 1981, **46**, 2861-2867.
152. O. Abril and G. M. Whitesides, *J. Am. Chem. Soc.*, 1982, **104**, 1552-1554.
153. R. Benhamza, Y. Amrani, and D. Sinou, *J. Organomet. Chem.*, 1985, **288**, C37-C39.
154. V. Ravindar, H. Hemling, H. Schumann, and J. Blum, *Synth. Commun.*, 1992, **22**, 841-851.
155. M. J. H. Russell, *Platinum Met. Rev.*, 1988, **32**, 179-186.
156. G. P. Schiemenz, *Chem. Ber.*, 1966, **99**, 504-513.
157. M. J. H. Russell and A. B. Murrer, Johnson Matthey Public Ltd. Co., *DE 3 135 127*, 1982; *GB 2 085 874*, 1981.
158. M. J. H. Russell and A. B. Murrer, Johnson Matthey PLC., *BE 890 210*, 1982; *Chem. Abstr.*, 1982, **97**, 23291n.
159. M. J. H. Russell, Johnson Matthey PLC., *FR 2 489 308*, 1982; *Chem. Abstr.*, 1982, **97**, 55308q.
160. H. Gilman and G. E. Brown, *J. Am. Chem. Soc.*, 1945, **67**, 824-826.
161. V. Ravindar, H. Hemling, H. Schumann, and J. Blum, *Synth. Commun.*, 1992, **22**, 1453-1459.
162. A. Badener and E. Lindner, *Coord. Chem. Rev.*, 1991, **108**, 27-110.
163. R. Luckenbach and K. Lorenz, *Z. Naturforsch. B*, 1977, **32**, 1038-1047.
164. O. Herd, A. Hessler, M. Hingst, M. Tepper, and O. Stelzer, *J. Organomet. Chem.*, 1996, **522**, 69-76.
165. A. Buhling, P. C. J. Kamer, and P. W. N. M. van Leeuwen, *J. Mol. Catal. A: Chem.*, 1995, **98**, 69-80.
166. F. Joó, L. Nádasdi, A. Benyei, P. Csiba and A. Kathó, in Ref. 70, p. 23-32.
167. F. G. Mann and I. T. Millar, *J. Chem. Soc.*, 1952, 4453-4457.
168. L. D. Pettit and H. M. N. H. Irving, *J. Chem. Soc.*, 1964, 5336-5343.

169. Y. Dror and J. Manassen, *Stud. Surf. Sci. Catal.*, 1981, **7**(Pt. B, New Horiz. Catal.), 887-897.
170. J. Podlahová and J. Podlaha, *Collect. Czech. Chem. Commun.*, 1980, **45**, 2049-2053.
171. A. Jegorov and J. Podlaha, *Catal. Lett.*, 1991, **9**, 9-14.
172. L. Lecomte, J. Triolet, D. Sinou, J. Bakos, and B. Heil, *J. Chromatogr.*, 1987, **408**, 416-419.
173. L. Lecomte and D. Sinou, *J. Chromatogr.*, 1990, **514**, 91-96.
174. J. Podlahová, B. Kratochvil, V. Langer, J. Silha, and J. Podlaha, *Collect. Czech. Chem. Commun.*, 1981, **46**, 3063-7073.
175. J. Podlaha and J. Podlahová, *Collect. Czech. Chem. Commun.*, 1973, **38**, 1730-1736.
176. J. Podlahová, *Collect. Czech. Chem. Commun.*, 1978, **43**, 3007.
177. M. M. Rauhut, I. Hechenbleikner, H. A. Currier, F. C. Schaffner, and V. P. Wystrach, *J. Am. Chem. Soc.*, 1959, **81**, 1103.
178. F. Hartl and J. Podlahová, *Collect. Czech. Chem. Commun.*, 1984, **49**, 586.
179. D. Nosková and J. Podlahová, *Polyhedron*, 1983, **2**, 349.
180. J. Podlahová, *Collect. Czech. Chem. Commun.*, 1978, **43**, 57.
181. T. Jarolim and J. Podlahová, *J. Inorg. Nucl. Chem.*, 1976, **38**, 125.
182. J. Podlahová, B. Kratochvil, J. Podlaha, and J. Hasek, *J. Chem. Soc., Dalton Trans.*, 1985, 2393-2396.
183. H. D. Empsall, E. M. Hyde, D. Pawson, and B. L. Shaw, *J. Chem. Soc., Dalton Trans.*, 1977, 1292-1298.
184. J. Chatt, G. J. Leigh, and R. M. Slade, *J. Chem. Soc., Dalton Trans.*, 1973, 2021-2028.
185. F. Joó and E. Trócsányi, *J. Organomet. Chem.*, 1982, **231**, 63-70.
186. F. Mercier and F. Mathey, *J. Organomet. Chem.*, 1993, **462**, 103-106.
187. A. Avey, D. M. Schut, T. J. R. Weakley, and D. R. Tyler, *Inorg. Chem.*, 1993, **32**, 233-236.
188. T. Malmstrom, H. Weigl, and C. Andersson, *Organometallics*, 1995, **14**, 2593-2596.
189. T. Malmstrom and C. Andersson, *Chem. Commun.*, 1996, 1135-1136.
190. M. E. Wilson, R. G. Nuzzo, and G. M. Whitesides, *J. Am. Chem. Soc.*, 1978, **100**, 2269-2270.
191. I. Willner, R. Maidan, and M. Shapira, *J. Chem. Soc., Perkin Trans.*, 1990, 559-564.
192. K. N. Harrison, P. A. T. Hoye, A. G. Orpen, P. G. Pringle, and M. B. Smith, *J. Chem. Soc., Chem. Commun.*, 1989, 1096-1097.
193. J. W. Ellis, K. N. Harrison, P. A. T. Hoye, A. G. Orpen, P. G. Pringle, and M. B. Smith, *Inorg. Chem.*, 1992, **31**, 3026-3033.
194. P. G. Pringle, D. Brewin, M. B. Smith, and K. Worboys, in Ref. 70, p. 111-122.
195. P. G. Pringle and M. B. Smith, *Platinum Met. Rev.*, 1990, **34**, 74-76.
196. P. A. T. Hoye, P. G. Pringle, M. B. Smith, and K. Worboys, *J. Chem. Soc., Dalton Trans.*, 1993, 269-274.
197. N. W. Alcock, A. W. G. Platt, and P. G. Pringle, *J. Chem. Soc., Dalton Trans.*, 1989, 139-143.
198. V. S. Reddy, D. E. Berning, K. V. Katti, C. L. Barnes, W. A. Volkert, and A. R. Ketring, *Inorg. Chem.*, 1996, **35**, 1753-1757.
199. C. J. Smith, V. S. Reddy, and K. V. Katti, *Chem. Commun.*, 1996, 2557-2558.
200. Z. Jin, Y. Yan, H. Zuo, and B. Fell, *J. Prakt. Chem./Chem.-Ztg.*, 1996, **338**, 124-128.
201. Y. Y. Yan, H. P. Zhuo, B. Yang, and Z. L. Jin, *J. Nat. Gas Chem.*, 1994, **3**, 436-441.
202. Y. Y. Yan, H. P. Zuo, and Z. L. Jin, *Chin. Chem. Lett.*, 1996, **7**, 377-380.
203. Y. Yan, H. Zhuo, and Z. Jin, *Fenzi Cuihua*, 1994, **8**, 147-150; *Chem. Abstr.*, 1994, **121**, 111875a.
204. A. Tzschach and E. Nietzschmann, *Z. Chem.*, 1980, **20**, 341.
205. O. Neunhoeffer and L. Lamza, *Chem. Ber.*, 1961, **94**, 2514-2518.
206. D. E. Bergbreiter, L. Zhang, and V. M. Mariagnanam, *J. Am. Chem. Soc.*, 1993, **115**, 9295- 9296.
207. Y. Amrani and D. Sinou, *J. Mol. Catal.*, 1984, **24**, 231-233.
208. D. Sinou and Y. Amrani, *J. Mol. Catal.*, 1986, **36**, 319-327.

209. T. N. Mitchell and K. Heesche-Wagner, *J. Organomet. Chem.*, 1992, **436**, 43-53.
210. K. Issleib and G. Thomas, *Chem. Ber.*, 1960, **93**, 803-808.
211. K. Issleib and G. Thomas, *Chem. Ber.*, 1961, **94**, 2244-2251.
212. K. Issleib and H. M. Mobius, *Chem. Ber.*, 1961, **94**, 102-106.
213. K. Issleib and H. R. Roloff, *Chem. Ber.*, 1965, **98**, 2091-2098.
214. T. Okano in Ref. 70, p. 97-110.
215. T. Okano, Y. Moriyama, H. Konishi, and J. Kiji, *Chem. Lett.*, 1986, 1463-1466.
216. T. Okano, M. Iwahara, H. Konishi, and J. Kiji, *J. Organomet. Chem.*, 1988, **346**, 267-275.
217. R. T. Smith, R. K. Ungar, L. S. Sanderson, and M. C. Baird, *Organometallics*, 1983, **2**, 1138- 1144.
218. R. T. Smith and M. C. Baird, *Trans. Met. Chem.*, 1981, **6**, 197-198.
219. M. K. Markiewicz and M. C. Baird, *Inorg. Chim. Acta*, 1986, **113**, 95.
220. R. T. Smith and M. C. Baird, *Inorg. Chim. Acta*, 1982, **62**, 135-139.
221. F. Joó and A. Kathó, *J. Mol. Catal. A: Chem.*, 1997, **116**, 3-26.
222. E. A. Broger, R. Schmid, Y. Crameri, M. Lalonde, A. Meili, and G. Schoettel, in "Reprints NATO ARW on Aqueous Organometallic Chemistry and Catalysis", Debrecen, Hungary, 29 Aug.- 1 Sept. 1994.
223. 209th Meeting of the American Chemical Society, Anaheim, CA, 1995, Abstracts of papers.
224. B. Cornils, W. A. Herrmann, and R. W. Eckel, *J. Mol. Catal. A: Chem.*, 1997, **116**, 27-33.
225. I. T. Horváth, Ed. in special issue on "Catalysis in Water", *J. Mol. Catal. A: Chem.*, 1997, **116**, p. 1-316.
226. M. Y. Darensbourg, ed. *Inorg. Synth.*, 1997, **32**, in press.
227. F. Bitterer, S. Kucken, and O. Stelzer, *Chem. Ber.* 1995, **128**, 275-279.
228. G. T. Baxley, T. J. R. Weakley, W. K. Miller, D. K. Lyon, and D. R. Tyler, *J. Mol. Catal. A: Chem.*, 1997, **116**, 191-198.
229. D. C. Mudalige and G. L. Rempel, *J. Mol. Catal. A: Chem.*, 1997, **116**, 309-316.
230. D. J. Brauer, J. Fischer, S. Kucken, K. P. Langhans, O. Stelzer, and N. Weferling, *Z. Naturforsch., B: Chem. Sci.*, 1994, **49**, 1511-1524.
231. A. Hessler, S. Kucken, O. Stelzer, J. Blotevogel-Baltronat, and W. S. Sheldrick, *J. Organomet. Chem.*, 1995, **501**, 293-302.
232. G. Peiffer, S. Chhan, A. Bendayan, B. Waegell, and J. P. Zahra, *J. Mol. Catal.*, 1990, **59**, 1-9.
233. H. Dibowski and F. P. Schmidtchen, *Tetrahedron*, 1995, **51**, 2325-2330.
234. G. Papadogianakis, in Ref. 28, p. 139-142.
235. E. Konigs and H. Friedrich, *Lieb. Ann. Chem.*, 1934, **509**, 138.
236. G. Tomaschewski, *J. Prakt. Chem.*, 1966, **33**, 168.
237. I. Tóth, B. E. Hanson, and M. E. Davis, *Organometallics*, 1990, **9**, 675.
238. A. Buhling, P. C. J. Kamer, P. W. N. M. van Leeuwen, and J. W. Elgersma, *J. Mol. Catal. A: Chem.*, 1997, **116**, 297-308.
239. I. Tóth and B. E. Hanson, *Tetrahedron Asymmetry*, 1990, **1**, 895-912.
240. I. Tóth, B. E. Hanson, and M. E. Davis, *Tetrahedron Assymetry*, 1990, **1**, 913-930.
241. I. Tóth, B. E. Hanson, and M. E. Davis, *J. Organomet. Chem.*, 1990, **396**, 363.
242. I. Tóth, B. E. Hanson, and M. E. Davis, *J. Organomet. Chem.*, 1990, **396**, 109.
243. I. Tóth, B. E. Hanson, and M. E. Davis, *Catal. Lett.*, 1990, **5**, 183.
244. D. J. Daigle, A. B. Pepperman, and S. L. Vail, *J. Heterocyclic Chem.*, 1974, **11**, 407.
245. D. J. Darensbourg, T. J. Decuir, and J. H. Reibenspies, in Ref. 70, p. 61-80.
246. D. R. Tyler, in Ref. 70, p. 47-60.
247. D. J. Darensbourg, F. Joó, M. Kannisto, A. Kathó, and J. H. Reibenspies, *Organometallics*, 1992, **11**, 1990-1993.
248. D. J. Darensbourg, N. W. Stafford, F. Joó, and J. H. Reibenspies, *J. Organomet. Chem.*, 1995, **488**, 99-108.
249. M. Y. Darensbourg and D. J. Daigle, *Inorg. Chem.*, 1975, **14**, 1217.

250. D. M. Saysell, C. D. Borman, C. H. Kwak, and A. G. Sykes, *Inorg. Chem.*, 1996, **35**, 173- 178.
251. Z. Assefa, B. G. McBurnett, R. J. Staples, and J. P. Fackler, *Inorg. Chem.*, 1995, **34**, 4965- 4972.
252. J. M. Forward, Z. Assefa, and J. P. Fackler, *J. Am. Chem. Soc.*, 1995, **117**, 9103-9104.
253. Z. Assefa, B. G. McBurnett, R. J. Staples, J. P. Fackler, B. Assmann, K. Angermeier, and H. Schmidbaur, *Inorg. Chem.*, 1995, **34**, 75-83.
254. B. Assmann, K. Angermeier, and H. Schmidbaur, *J. Chem. Soc., Chem. Commun.*, 1994, 941- 942.
255. E. Renaud, R. B. Russell, S. Fortier, S. J. Brown, and M. C. Baird, *J. Organomet. Chem.*, 1991, **419**, 403-415.
256. E. Renaud and M. C. Baird, *J. Chem. Soc., Dalton Trans.*, 1992, 2905.
257. I. Kovacs and M. C. Baird, *J. Organomet. Chem.*, 1995, **502**, 87-94.
258. T. L. Schull, J. C. Fettinger, and D. A. Knight, *J. Chem. Soc., Chem. Commun.*, 1995, 1487- 1488.
259. T. L. Schull, J. C. Fettinger, and D. A. Knight, *Inorg. Chem.*, 1996, **35**, 6717-6723.
260. S. Levievre, F. Mercier, and F. Mathey, *J. Org. Chem.*, 1996, **61**, 3531-3533.
261. D. J. Darensbourg, F. Joó, M. Kannisto, A. Kathó, J. R. Reibenspies, and D. J. Daigle, *Inorg. Chem.*, 1994, **33**, 200-208.
262. F. Farin, H. L. M. van Gaal, S. L. Bontinger, and F. J. M. Daemen, *Biochim. Biophys. Acta*, 1982, **711**, 336-344.
263. Z. Jiang and A. Sen, *Macromolecules*, 1994, **27**, 7215-7216.
264. G. Schmid, B. Morun, and J. O. Malm, *Angew. Chem.*, 1989, **101**, 772-773; *Angew. Chem. Int. Ed. Engl.*, 1989, **28**, 778-780.
265. G. Schmid and H. Hess, *Z. Anorg. Allg. Chem.*, 1995, **621**, 1147-1152.
266. G. Schmid, in "Clusters and Colloids: From Theory to Applications", ed. G. Schmid, VCH, Weinheim, 1994, p. 182.
267. W. A. Herrmann, W. R. Thiel, and J. G. Kuchler, *Chem. Ber.*, 1990, **123**, 1953-1961.
268. W. A. Herrmann, W. R. Thiel, J. G. Kuchler, J. Behm, and E. Herdtweck, *Chem. Ber.*, 1990, **123**, 1963-1970.
269. S. Anderson, E. C. Constable, K. R. Seddon, H. E. Turp, J. E. Baggot, and M. J. Pilling, *J. Chem. Soc., Dalton Trans.*, 1985, 2247.
270. G. Sprintschnik, H. W. Sprintschnik, P. P. Kirsch, and D. G. Whitten, *J. Am. Chem. Soc.*, 1977, **99**, 4947-4954.
271. H. Irving, R. Shelton, and R. Evans, *J. Chem. Soc.*, 1958, 3540.
272. S. Kolaric and V. Sunjic, *J. Mol. Catal. A: Chem.*, 1996, **110**, 181-188.
273. W. S. Knowles, W. C. Christopfel, K. E. Koenig, and C. F. Hobbs, in "Catalytic Aspects of Metal Phosphine Complexes" eds. E. C. Alyea and D. W. Meek, Advances in Chemistry Series, American Chemical Society, Washington, D. C., 1982, No. 196, p. 325-336.
274. L. Vigh and F. Joó, in Ref. 70, p. 281-287.
275. F. Joó, N. Balogh, L. I. Horváth, G. Filep, I. Horváth, and L. Vigh, *Anal. Biochem.*, 1991, **194**, 34-40.
276. P. Scrimin and U. Tonellato, in "Surfactants in Solution", ed. K. L. Mittal and D. O. Shah, Plenum Press, New York, 1991, vol. 11, p. 349-362.
277. R. Fornasier, P. Scrimin, P. Tecilla, and U. Tonellato, *J. Am. Chem. Soc.*, 1989, **111**, 224- 229.
278. K. Ogino, Y. Tokuda, T. Nakai, and W. Tagaki, *Mem. Fac. Eng., Osaka City Univ.*, 1994, **35**, 187-198.
279. W. Tagaki, K. Ogino, T. Fujita, T. Yoshida, K. Nishi, and Y. Inaba, *Bull. Chem. Soc. Jpn.*, 1993, **66**, 140-147.
280. J. H. Weber and D. H. Busch, *Inorg. Chem.*, 1965, **4**, 469-471.
281. J. H. Weber and D. H. Busch, *Inorg. Chem.*, 1965, **4**, 472-475.
282. C. Bianchini, P. Frediani, and V. Sernau, *Organometallics*, 1995, **14**, 5458-5459.
283. B. Fell, G. Papadogianakis, W. Konkol, J. Weber, and H. Bahrmann, *J. Prakt. Chem./Chem.- Ztg.*, 1993, **335**, 75-82.

284. H. Bahrmann, B. Fell, and G. Papadogianakis, Hoechst AG, *EP 435 071*, 1990, *DE 3 942 787*, 1989; *Chem. Abstr.*, 1991, **115**, 183583c.
285. H. Bahrmann, B. Fell, and G. Papadogianakis, Hoechst AG, *EP 435 084*, 1990, *DE 3 942 954*, 1989; *Chem. Abstr.*, 1991, **115**, 255627v.
286. G. Papadogianakis, in Ref. 28, p. 177-208.
287. H. Bahrmann, in Ref. 31, p. 644-654.
288. T. Okano, N. Harada, and J. Kiji, *Chem. Lett.*, 1994, 1057.
289. A. Hercouet and M. Le Corre, *Synthesis*, 1988, 157-158.
290. T. S. Srivastava and M. Tsutsui, *J. Org. Chem.*, 1973, **38**, 2103.
291. E. B. Fleischer, J. M. Palmer, T. S. Srivastava, and A. Chatterjee, *J. Am. Chem. Soc.*, 1971, **93**, 3162-3167.
292. J. A. Peters, J. Huskens, and D. J. Raber, *Prog. Nucl. Magn. Reson. Spectrosc.*, 1996, **28**, 283-350.
293. R. D. Hancock and A. E. Martell, *Chem. Rev.*, 1989, **89**, 1875-1914.
294. V. Alexander, *Chem. Rev.*, 1995, **95**, 273-342.
295. Pharmacyclics Inc., in "Texaphyrins: An Overview", www: http://www.pcyc.com/Website/ TechDocs/texaphrn.htm (22.05.1997).
296. W. D. W. Horrocks and E. G. Hove, *J. Am. Chem. Soc.*, 1978, **100**, 4386-4392.
297. A. Treibs and N. Häberle, *Liebigs Ann. Chem.*, 1968, **718**, 183-207.
298. M. Pawlik, M. F. Hoq, and R. E. Shepherd, *J. Chem. Soc., Chem. Commun.*, 1983, 1467- 1468.
299. P. C. Ford and A. Rokicki, *Adv. Organomet. Chem.*, 1988, **28**, 139-217.
300. J. Kido, Y. Okamoto, and H. G. Brittain, *J. Org. Chem.*, 1991, **56**, 1412-1415.
301. R. Hulst, N. K. de Vries, and B. L. Feringa, *J. Org. Chem.*, 1994, **59**, 7453-7458.
302. A. D. Sherry, G. R. Malloy, F. M. H. Jeffrey, W. P. Cacheris, and C. F. G. C. Geraldes, *J. Magn. Reson.*, 1988, **76**, 528-533.
303. D. W. Swinkels, J. P. M. van Duynhoven, C. W. Hilbers, and G. I. Tesser, *Recl. Trav. Chim. Pays-Bas*, 1991, **110**, 124-128.
304. C. F. G. C. Geraldes, A. D. Sherry, and W. P. Cacheris, *Inorg. Chem.*, 1989, **28**, 3336-3341.
305. G. Papadogianakis, L. Maat, and R. A. Sheldon, *J. Mol. Catal. A: Chem.*, 1997, **116**, 179-190.
306. G. Papadogianakis and B. Fell, 1987, unpublished results.
307. Th. Müller and H. Bahrmann, *J. Mol. Catal. A: Chem.*, 1997, **116**, 39-42.
308. O. J. Gelling, I. Toth, DSM N.V., E. I. du Pont de Nemours and Co., *WO 95/18783*, 1995; *Chem. Abstr.*, 1995, **123**, 285242y.
309. G. Oehme, E. Paetzold, and A. Kinting, Akademie der Wissenschaften, *DD 259 194*, 1988; *Chem. Abstr.*, 1989, **111**, 58020h.
310. O. Roelen, Ruhrchemie AG/Chemische Verwertungsgesellschaft Oberhausen m.b.H., *D.B.P. 849 548*, 1938/1952; *Chem. Zbl.*, 1953, **124**, 927.
311. O. Roelen, *Angew. Chem. A*, 1948, **60**, 213.
312. O. Roelen, *Angew. Chem.*, 1951, **63**, 482-483.
313. B. Cornils, W. A. Herrmann, and M. Rasch, *Angew. Chem.*, 1994, **106**, 2219-2238; *Angew. Chem. Int. Ed. Engl.*, 1994, **33**, 2144-2163.
314. O. Roelen, in "Präparative Organische Chemie", ed. K. Ziegler, Dieterich'sche Verlagsbuchandlung, Wiesbaden, 1948, **36**, Part I, B, 155-170.
315. H. Adkins and G. Krsek, *J. Am. Chem. Soc.*, 1948, **70**, 383-386.
316. H. Adkins and G. Krsek, *J. Am. Chem. Soc.*, 1949, **71**, 3051-3055.
317. H. Turk and W. T. Ford, *J. Org. Chem.*, 1988, **53**, 460-462.
318. W. M. Brouwer, P. Piet, and A. L. German, *J. Mol. Catal.*, 1985, **31**, 169-182.
319. R. A. Sheldon and J. Dakka, *Catal. Today*, 1994, **19**, 215-246.
320. A. Buhling, Ph. D. Thesis, Universiteit van Amsterdam, The Netherlands, 1996.
321. B. Cornils, in "New Syntheses with Carbon Monoxide", ed. J. Falbe, Springer Verlag, Berlin, 1980, ch.1, p. 1-225.
322. H. Bach, W. Gick, E. Wiebus, and B. Cornils, *Proc. 8th Int. Congr. Catal.*, Verlag Chemie, Weinheim, 1984, **5**, V417-V427.

323. H. Bach, W. Gick, W. Konkol, and E. Wiebus, *Proc. 9th Int. Congr. Catal.*, ed. M. J. Phillips and M. Ternan, 1988, **1**, 254-259.
324. L. H. Slaugh and R. D. Mullineaux, Shell Oil Co., *U.S. 32 39 569*, 1960.
325. L. H. Slaugh and R. D. Mullineaux, Shell Oil Co., *U.S. 32 39 570*, 1960.
326. D. Evans, J. A. Osborn, and G. Wilkinson, *J. Chem. Soc., A*, 1968, 31.
327. J. L. Paul, W. L. Pieper, and L. E. Wade, Celanese Corp., *DE 21 25 382*, 1971.
328. A. L. Stautzenberger and J. L. Paul, Celanese Corp., *U.S. 40 09 003*, 1977.
329. R. L. Pruett, Union Carbide Corp., *DE 17 93 069*, 1972.
330. Anonymous, (Heart Cut) *Chemtech*, 1978, **8**, 8 and 43.
331. H. Haas, Böhme Fettchemie GmbH, *DE 937 949*, 1956; *Chem. Abstr.*, 1956, **50**, 12108; *Chem. Zbl.*, 1957, **128**, 2049.
332. R. Hage, J. E. Iburg, J. Kerschner, J. H. Koek, E. L. M. Lempers, R. J. Martens, U. S. Racherla, S. W. Russell, T. Swarthoff, M. R. P. v. Vliet, J. B. Warnaar, L. v.d. Wolf, and B. Krijnen, *Nature (London)*, 1994, **369**, 637-639.
333. E. Paetzold, M. Michalik, and G. Oehme, *J. Prakt. Chem./Chem.-Ztg.*, 1997, **339**, 38-43.
334. Z. Xia and B. Fell, *J. Prakt. Chem./Chem.-Ztg.*, 1997, **339**, 140-144.
335. J. A. van Doorn, E. Drent, P. W. N. M. van Leeuwen, N. Meijboom, A. B. van Oort, and R. L. Wife, Shell Int. Res. Maat. B. V., *EP 280 380*, 1988; *Chem. Abstr.*, 1989, **110**, 58296g.
336. H. Bach, H. Bahrmann, W. Gick, W. Konkol, and E. Wiebus, *Chem. Ing. Tech.*, 1987, **59**, 882-883.
337. H. Bahrmann and H. Bach, *Phosphorus Sulfur*, 1987, **30**, 611-614.
338. G. Kessen and B. Cornils, Ruhrchemie AG, *DE 35 30 839*, 1987; *Chem. Abstr.*, 1987, **107**, 42047k.
339. Ruhrchemie AG, *CN 85 105 099*, 1985; *Chem. Abstr.*, 1888, **108**, 96632e.
340. B. Cornils, W. Konkol, and H. Bahrmann, Ruhrchemie AG, *DE 35 46 123* , 1987; *Chem. Abstr.*, 1987, **107**, 153922e.
341. B. Cornils, W. Konkol, H. Bach, G. Daemkes, W. Gick, E. Wiebus, and H. Bahrmann, Ruhrchemie AG, *DE 34 15 968*; 1984; *Chem. Abstr.*, 1986, **104**, 209147p.
342. B. Cornils, W. Konkol, H. Bach, G. Daemkes, W. Gick, W. Greb, E. Wiebus, and H. Bahrmann, Ruhrchemie AG, *DE 34 13 427*, 1984; *Chem. Abstr.*, 1986, **104**, 188661j.
343. H. Kalbfell, B. Lieder, and H. Mercamp, Ruhrchemie AG, *DE 33 41 035*, 1983; *Chem. Abstr.*, 1986, **104**, 19340y.
344. H. Bahrmann, H. Bach, C. D. Frohning, H. J. Kleiner, P. Lappe, D. Peters, D. Regnat, and W. A. Herrmann, *J. Mol. Catal. A: Chem.*, 1997, **116**, 49-53.
345. I. T. Horváth, R. v. Kastrup, A. A. Oswald, and E. J. Mozeleski, *Catal. Lett.*, 1989, **2**, 85-90.
346. M. J. H. Russell, *Chem.-Tech.* (Heidelberg) 1988 (Int. Issue) 17-22.
347. M. J. H. Russell, *Spec. Chem.*, 1988, **8**, 398-400.
348. Anonymous, *Eur. Chem. News*, 1995, Jan. 15, 29.
349. W. Greb, J. Hibbel, J. Much, and V. Schmidt, Ruhrchemie AG, *EP 263 953*, 1986.
350. H. Bahrmann, C. D. Frohning, P. Heymanns, H. Kalbfell, P. Lappe, D. Peters, and E. Wiebus, *J. Mol. Catal. A: Chem.*, 1997, **116**, 35-37.
351. W. Keim, A. Behr, and G. Schmitt, "Grundlagen der Industriellen Chemie", Otto Salle Verlag, Verlag Sauerländer, Frankfurt am Main, 1986, p.129-133.
352. B. Fell and P. Hermanns, Hoechst AG, *EP 643 031*, 1995; *Chem. Abstr.*, 1995, **122**, 264902c.
353. H. Bahrmann, B. Cornils, W. Konkol, and W. Lipps, Ruhrchemie AG, *DE 3 420 491*, 1985; *Chem. Abstr.*, 1986, **105**, 42477c.
354. N. Calamur, M. E. Carrera, and R. A. Wilsak, in "Kirk-Othmer: Encyclopedia of Chemical Technology", ed. J. I. Kroschwitz and M. Howe-Grant, Wiley, New York, 4th edn, 1992, vol. 4, p. 701-735.

355. K. Weissermel and H. J. Arpe, "Industrial Organic Chemistry", VCH, Weinheim, 2nd edn, 1993, ch.3, p.59-88.
356. E. Billig, in Ref. 354, p. 736-747.
357. K. Kurtev, D. Ribola, R. A. Jones, D. J. Cole-Hamilton, and G. Wilkinson, *J. Chem. Soc., Dalton Trans.*, 1980, 55-58.
358. R. T. Smith, Ph.D. Thesis, Queen's University at Kingston, Canada, *Diss. Abstr. Int. B*, 1982, **43**, 1499; *Chem. Abstr.*, 1983, **98**, 72414j.
359. P. Escaffre, A. Thorez, and Ph. Kalck, *J. Chem. Soc., Chem. Commun.*, 1987, 146-147.
360. B. Cornils, H. Bahrmann, W. Lipps, and W. Konkol, Ruhrchemie AG, *EP 173 219*, 1986; *Chem. Abstr.*, 1986, **105**, 81168x.
361. B. Cornils, H. Bahrmann, W. Lipps, and W. Konkol, Ruhrchemie AG, *DE 3 511 428*, 1986; *Chem. Abstr.*, 1987, **106**, 521059.
362. Ph. Kalck, *Polyhedron*, 1988, **7**, 2441-2450.
363. R. V. Chaudhari, B. M. Bhanage, R. M. Deshpande, and H. Delmas, *Nature (London)*, 1995, **373**, 501-503.
364. R. V. Chaudhari, B. M. Bhanage, S. S. Divekar, and R. M. Deshpande, Council of Scientific & Industrial Interest, *US 54 98 801*, 1996; *Chem. Abstr.*, 1996, **124**, 342663h.
365. P. Purwanto and H. Delmas, *Catal. Today*, 1995, **24**, 135-140.
366. R. M. Deshpande, P. Purwanto, H. Delmas, and R. V. Chaudhari, *Ind. Eng. Chem. Res.*, 1996, **35**, 3927-3933.
367. F. Monteil, R. Queau, and Ph. Kalck, *J. Organomet. Chem.*, 1994, **480**, 177-184.
368. F. Baumann and B. Bienert, Farbenfabriken Bayer, *US 2 613 128*, 1952.
369. G. Gregorio and A. Andreetta, Montecatini Edison S.p.A, *DE 23 13 102*, 1973.
370. O. J. Gelling and I. Toth, DSM N.V., *WO 95/19331*, 1995; *Chem. Abstr.*, 1995, **123**, 285233w.
371. E. A. Karakhanov, Y. S. Kardasheva, A. L. Maksimov, V. V. Predeina, E. A. Runova, and A. M. Utukin, *J. Mol. Catal. A: Chem.*, 1996, **107**, 235-240.
372. S. J. McLain, *J. Am. Chem. Soc.*, 1983, **105**, 6355-6357.
373. G. Papadogianakis, in Ref. 28, p. 62 and 96.
374. H. Bahrmann, W. A. Herrmann, C. Kohlpaintner, and R. Manetsberger, Hoechst AG, *DE 4 220 267*, 1993; *Chem. Abstr.*, 1994, **120**, 218184f.
375. W. Herrmann , C. Kohlpaintner, and H. Bahrmann, Hoechst AG, *EP 491 240*, 1992; *Chem. Abstr.*, 1992, **117**, 151154p.
376. T. Allman and R. G. Goel, *Can. J. Chem.*, 1982, **60**, 716-722.
377. J. D. Unruh and J. R. Christenson, *J. Mol. Catal.*, 1982, **14**, 19-34.
378. D. W. Meek, in "Homogeneous Catalysis with Metal Phosphine Complexes", ed. L. H. Pignolet, Modern Inorganic Chemistry, Ser. ed. J. P. Fackler, Plenum, New York, 1983, ch.8, p.257-296.
379. W. R. Moser, C. J. Papile, D. A. Brannon, R. A. Duwell, and S. J. Weininger, *J. Mol. Catal.*, 1987, **41**, 271-292.
380. H. Bach, H. Bahrmann, B. Cornils, W. Konkol, and E. Wiebus, Ruhrchemie AG, *DE 3 616 057*, 1987; *Chem. Abstr.*, 1988, **108**, 152528q.
381. B. Cornils, W. Konkol, H. Bahrmann, H. Bach, and E. Wiebus, Ruhrchemie AG, *DE 3 443 474*, 1986; *Chem. Abstr.*, 1986, **105**, 210764q.
382. B. Cornils, W. Konkol, H. Bahrmann, H. Bach, and E. Wiebus, Ruhrchemie AG, *DE 3 411 034*, 1985; *Chem. Abstr.*, 1986, **105**, 97007e.
383. R. Gärtner, B. Cornils, L. Bexten, and D. Kupies, Ruhrchemie AG, *DE 3 235 029*, 1984; *Chem. Abstr.*, 1984, **100**, 198449c.
384. G. Daembkes, H. D. Hahn, J. Hibbel, and W. Materne, Ruhrchemie AG, *DE 3 347 406*, 1985; *Chem. Abstr.*, 1985, **103**, 127031m.
385. B. Cornils, J. Hibbel, W. Konkol, B. Lieder, J. Much, V. Schmidt, and E. Wiebus, Ruhrchemie AG, *DE 3 234 701*, 1984; *Chem. Abstr.*, 1984, **100**, 194022k.
386. E. J. R. Sudhölter, *Polytech. Tijdschr. Procestech.*, 1978, **33**, 443-446.
387. W. C. Griffin, in "Kirk-Othmer: Encyclopedia of Chemical Technology", ed, H. F.

Mark, D. F. Othmer, C. G. Overberger and G. T. Seaborg, Wiley, New York, 3rd edn, 1979, vol. 8, p. 900-930.
388. J. T. Davies, *Proc. 2nd Int. Congr. Surf. Act. I*, 1957, 426-438.
389. J. H. Fendler and E. J. Fendler, "Catalysis in Micellar and Macromolecular Systems", Academic Press, New York, 1975.
390. E. J. Fendler and J. H. Fendler, *Adv. Phys. Org. Chem.*, 1970, **8**, 271-406.
391. E. H. Cordes and C. Gitler, *Prog. Bioorg. Chem.*, 1973, **2**, 1-53.
392. E. H. Cordes and R. B. Dunlap, *Acc. Chem. Res.*, 1969, **2**, 329-337.
393. L. H. Slaugh and R. D. Mullineaux, *J. Organomet. Chem.*, 1968, **13**, 469.
394. Ruhrchemie AG, *CN 85 105 102*, 1986; *Chem. Abstr.*, 1988, **108**, 133848c.
395. H. Bahrmann, B. Cornils, W. Konkol, and W. Lipps, Ruhrchemie AG, *DE 34 12 335*, 1985; *Chem. Abstr.*, 1986, **104**, 185989m.
396. Z. Jin, X. Zheng, and B. Fell, *J. Mol. Catal. A: Chem.*, 1997, **116**, 55-58.
397. H. Lange, *Fette Seifen Anstrichmittel*, 1968, **70**, 748.
398. N. L. Ferguson, *J. Am. Chem. Soc.*, 1955, **77**, 5288-5289.
399. L. Greenwald, G. L. Brown, *J. Phys. Chem.*, 1954, **58**, 825-828.
400. F. E. Bailey and J. V. Koleske, "Poly(ethylene oxide)", Academic Press, New York, 1976.
401. C. Bianchini, A. Meli, M. Peruzzini, F. Vizza, P. Frediani, and J. A. Ramirez, *Organometallics*, 1990, **9**, 226.
402. W. C. Griffin, *J. Soc. Cosmet. Chem.*, 1954, **5**, 249-256.
403. W. C. Griffin, *J. Soc. Cosmet. Chem.*, 1949, **1**, 311-326.
404. H. Bahrmann, W. Konkol, J. Weber, H. Bach, and L. Bexten, Hoechst AG, *DE 35 34 314*, 1987; *EP 216 315*, 1989; *Chem. Abstr.*, 1987, **107**, 42078q.
405. H. Bahrmann, B. Cornils, W. Konkol, J. Weber, L. Bexten, and H. Bach, Ruhrchemie AG, *DE 35 34 317*, 1985.
406. S. Riva, in "Applied Biocatalysis", ed. H. W. Blanch and D. S. Clark, Dekker, New York, 1991, vol. 1, ch. 5, p. 205.
407. B. C. Gates, "Catalytic Chemistry", Wiley, New York, 1992, (a) p. 96; (b) p.129-130.
408. A. G. Abatjoglou, D. R. Bryant, and R. R. Peterson, Union Carbide Chemicals and Plastics Co., Inc., *EP 350 922*, 1990; *US 218 911*, 1988; *US 5 180 854*, 1993; *Chem. Abstr*, 1990, **113**, 80944a.
409. A. G. Abatjoglou, R. R. Peterson, and D. R. Bryant, in "Catalysis of Organic Reactions", ed. R.E. Malz, Dekker, New York, 1996, p.133-139.
410. A. G. Abatjoglou and D. R. Bryant, Union Carbide Chemicals and Plastics Co., Inc., *EP 355 837*, 1990; *US 5 114 473*, 1992; *Chem. Abstr*, 1990, **113**, 9861r.
411. J. Haggin, *Chem. Eng. News*, 1995, **73**(Nr.16), 25-26.
412. M. J. H. Russell, *Platinum Met. Rev.*, 1997, **41**, 22-23.
413. A. G. Abatjoglou and D. R. Bryant, Union Carbide Corp., *US 4 731 486*, 1988; *Chem. Abstr*, 1988, **109**, 57042p.
414. L. Tinucci and E. Platone, Eniricerche S.p.A., *EP 380 154*, 1990; *Chem. Abstr.*, 1990, **113**, 214259s.
415. G. W. Parshall and S. D. Ittel, in "Homogeneous Catalysis: The Applications and Chemistry of Catalysis by Soluble Transition Metal Complexes", Wiley, New York, 2nd edn, 1992, p.111.
416. H. Bahrmann and P. Lappe, Hoechst AG, *EP 602 463*, 1994; *Chem. Abstr.*, 1994, **121**, 107982r.
417. B. Fell, Ch. Schobben, and G. Papadogianakis, *J. Mol. Catal. A: Chem.*, 1995, **101**, 179-186.
418. B. C. Gates, in "Kirk-Othmer: Encyclopedia of Chemical Technology" ed., J. I. Kroschwitz and M. Howe-Grant, Wiley, New York, 1993, 4th edn, vol. 5, p.336-337.
419. H. Bahrmann, P. Lappe, B. Fell, D. Leckel, and Ch. Schobben, Hoechst AG, *EP 711 748*, 1996; *Chem. Abstr.*, 1996, **125**, 89586.
420. E. N. Frankel, S. Metlin, W. K. Rohwedder and I. Wender, *J. Am Oil Chem. Soc.*, 1969, **46**, 133-138.

421. E. H. Pryde, *J. Am. Oil Chem. Soc.*, 1984, **61**, 419-425.
422. J. A. Cusumano, in "Perspectives in Catalysis: A Chemistry for the 21st Century monograph", ed. J. M. Thomas and K. I. Zamaraev, IUPAC, Blackwell, London, 1992, p.1-33.
423. G. Fremy, E. Monflier, J. F. Carpentier, Y. Castanet and A. Mortreux, *Angew. Chem.*, 1995, **107**, 1608-1610; *Angew. Chem. Int. Ed. Engl.*, 1995, **34**, 1474-1476.
424. B. L. Tarmy, in Ref. 418, vol. 20, p.1007-1059.
425. T. Omatsu and Y. Tokito, Kuraray Co., *JP 7 267 890*, 1995; *Chem. Abstr.*, 1996, **124**, 116665d.
426. T. Omatsu and Y. Tokito, Kuraray Co., *JP 7 267 891*, 1995; *Chem. Abstr.*, 1996, **124**, 116666e.
427. N. Yoshimura and M. Tamura, in "Successful Design of Catalysts'" ed. T. Inui, Studies in Surface Science and Catalysis, ed. B. Delmon and J. T. Yates, Elsevier, Amsterdam, 1989, vol. 44, p. 307-314.
428. N. Yoshimura and M. Tamura, Kuraray Co., *FR 2 499 978*, 1982; *US 4 417 079*, 1983; *Chem. Abstr.*, 1983, **98**, 71478w.
429. N. Yoshimura and M. Tamura, Kuraray Co., *FR 2 479 187*, 1981; *US 4 356 333*, 1982; *Chem. Abstr.*, 1982, **96**, 103630s.
430. T. Maeda, Y. Tokitoh, and N. Yoshimura, Kuraray Co., *EP 296 550*, 1988; *US 4 927 960*, 1990; *Chem. Abstr.*, 1989, **110**, 173469c.
431. G. W. Parshall, in Ref. 11, p.82-85.
432. A. Höhn, in Ref. 31, vol. 1, ch. 2, p. 138-147.
433. M. Beller and A. Tafesh, in Ref. 31, vol. 1, ch. 2, p. 187-200.
434. P. Hofmann, K. Kosswig, and W. Schaefer, *Ind. Eng. Chem. Prod. Res. Dev.*, 1980, **19**, 330- 334.
435. R. F. Heck, "Palladium Reagents in Organic Syntheses", Academic Press, London, 1985, p. 381-382.
436. J. F. Knifton, *J. Org. Chem.*, 1976, **41**, 2885-2890.
437. D. M. Fenton, *J. Org. Chem.*, 1973, **38**, 3192-3198.
438. D. E. James and J. K. Stille, *J. Am. Chem. Soc.*, 1976, **98**, 1810-1823.
439. W. P. Shum and J. F. White, Atlantic Richfield Co., *US 4 612 390*, 1986; *Chem. Abstr.*, 1987, **106**, 35085r; (cf. Example 8; run 125-22)
440. K. von Bittler, N. von Kutepow, D. Neubauer and H. Reis, *Angew. Chem.*, 1968, **80**, 352-359; *Angew. Chem. Int. Ed. Engl.*, 1968, **7**, 329.
441. E. N. Squire and F. J. Waller, E. I. du Pont de Nemours and Co., *US 4 292 437*, 1981; *Chem. Abstr.*, 1982, **96**, 6206a.
442. F. A. Pesa, T. A. Haase, Standard Oil Co., *US 4 469 886*, 1984.
443. F. A. Pesa, T. A. Haase, Standard Oil Co., *EP 52 419*, 1980.
444. T. C. Wu, Ethyl Corp., *US 5 315 028*, 1994.
445. G. Cavinato and L. Toniolo, *J. Mol. Catal.*, 1979, **6**, 111-122.
446. V. Elango, M. A. Murphy, B. L. Smith, K. G. Davenport, G. N. Mott, E. G. Zey, and G. L. Moss, Hoechst Celanese Corp., *U. S. 4 981 995*, 1991; *EP 400 892*, 1990; *Chem. Abstr.* 1991, **114**, 206780k.; *EP 284 310*, 1988; *Chem. Abstr.*, 1989 **110**, 153916t.
447. J. D. Hendricks and G. N. Mott, Hoechst Celanese Corp., *US 5 166 418*, 1992; *EP 460 905*, 1991; *Chem. Abstr.,*, 1992, **116**, 83378.
448. G. Papadogianakis, L. Maat and R. A. Sheldon, *Inorg. Synth.*, 1997, **32**, in press.
449. G. Papadogianakis, L. Maat, and R. A. Sheldon, *J. Chem. Soc., Chem. Commun.*, 1994, 2659- 2660.
450. G. Papadogianakis, J. A. Peters, L. Maat, and R. A. Sheldon, *J. Chem. Soc., Chem. Commun.*, 1995, 1105-1106.
451. G. Papadogianakis, L. Maat, and R. A. Sheldon, *J. Chem. Technol. Biotechnol.*, 1997, **68** in press.
452. G. Papadogianakis, G. Verspui, L. Maat, and R. A. Sheldon, *Catal. Lett.*, paper submitted.
453. V. V. Grushin, *Chem. Rev.*, 1996, **96**, 2011-2033.

454. D. Milstein, *J. Chem. Soc., Chem. Commun.*, 1986, 817-818.
455. L. F. Staroselskaya, T. E. Kron, and M. I. Terekhova, *Zh. Obshch. Khim.*, 1991, **61**, 736-739; *J. Gen. Chem. USSR (Engl. Transl.)*, 1991, **61**, 668-671.
456. B. F. M. Kuster, *Starch/Stärke*, 1990, **42**, 314-321.
457. M. Kunz, in "Inulin and Inulin-containing Crops", ed. A. Fuchs, Elsevier, Amsterdam, 1993, p. 149-160.
458. J. Haggin, *Chem. Eng. News*, 1995, **73**(Nr.7), 38.
459. K. G. Davenport, in "The Second International Symposium on Technologies for the Production of Enantiomerically Pure Chemicals", Naples, Florida, March 14-17, 1995.
460. J. N. Armor, *Appl. Catal.*, 1991, **78**, 141-173.
461. G. N. Mott and E. G. Zey, Hoechst Celanese Corp., *EP 337 803*, 1989; *Chem. Abstr.*, 1990, **112**, 234979.
462. R. A. Sheldon, L. Maat, and G. Papadogianakis, Hoechst Celanese Corp., *US 5 536 874*, 1996.
463. G. Cavinato and L. Toniolo, *J. Mol. Catal.*, 1993, **78**, 131
464. F. Monteil and Ph. Kalck, *J. Organomet. Chem.*, 1994, **482**, 45-51.
465. F. Monteil, L. Miquel, R. Queau, and Ph. Kalck, in Ref. 70, p.131-147.
466. C. W. Kohlpaintner and M. Beller, *J. Mol. Catal. A: Chem.*, 1997, **116**, 259-267.
467. C. W. Kohlpaintner and M. Beller, Hoechst AG, *DE 4 415 681*, 1995; *Chem. Abstr.*, 1996, **124**, 145625w.
468. C. W. Kohlpaintner and M. Beller, Hoechst AG, *DE 4 415 682*, 1995.
469. T. Okano, I. Uchida, T. Nakagaki, H. Konishi, and J. Kiji, *J. Mol. Catal.*, 1989, **54**, 65-71.
470. T. Okano, T. Hayashi, and J. Kiji, *Bull. Chem. Soc. Jpn.*, 1994, **67**, 2339-2341.
471. H. Jiang, Y. Xu, S. J. Liao, D. R. Yu, H. Chen, and X. J. Li, *Chin. Chem. Lett.*, 1996, **7**, 847- 848.
472. Z. Yu, Y. Xu, S. Liao, H. Jiang, B. Yang, D. Yu, H. Chen, and X. Li, *React. Funct. Polym.*, 1996, **31**, 201-205.
473. J. Kiji, T. Okano, W. Nishiumi, and H. Konishi, *Chem. Lett.*, 1988, 957-960.
474. T. Okano, N. Okabe, and J. Kiji, *Bull. Chem. Soc. Jpn.*, 1992, **65**, 2589-2593.
475. E. Monflier and A. Mortreux, *J. Mol. Catal.*, 1994, **88**, 295-300.
476. E. Monflier, A. Mortreux, and F. Petit, *Appl. Catal. A*, 1993, **102**, 53-67.
477. R. K. Freier, "Aqueous Solutions: Data for Inorganic and Organic Compounds", Walter de Gruyter, Berlin, 1976, vol. 1, p. 347.
478. F. M. Menger and C. E. Mounier, *J. Am. Chem. Soc.*, 1993, **115**, 12222-12223; and references therein.
479. I. Grassert, V. Vill and G. Oehme, *J. Mol. Catal. A: Chem.*, 1997, **116**, 231-236.
480. F. Joó and M. T. Beck, *React. Kinet. Catal. Lett.*, 1975, **2**, 257-263 (submitted in November 1974).
481. M. T. Beck, in Ref. 70, p. xv-xvi.
482. E. Fache, F. Senocq, C. Santini, and J. M. Basset, *J. Chem. Soc., Chem. Commun.*, 1990, 1776-1778.
483. E. Fache, C. Santini, F. Senocq, and J. M. Basset, *J. Mol. Catal.*, 1992, **72**, 331-336.
484. E. Fache, C. Santini, F. Senocq, and J. M. Basset, *J. Mol. Catal.*, 1992, **72**, 337-350.
485. J. M. Grosselin, C. Mercier, G. Allmang, and F. Grass, *Organometallics*, 1991, **10**, 2126- 2133.
486. J. M. Grosselin and C. Mercier, *J. Mol. Catal.*, 1990, **63**, L25-L27.
487. J. M. Grosselin, Rhône-Poulenc, *EP 03 62 037*, 1989; *Chem. Abstr.*, 1990, **113**, 77911g.
488. H. Brunner, in Ref.31, vol.1, ch.2, p.201-219.
489. G. Allmang, F. Grass, J. M. Grosselin and C. Mercier, *J. Mol. Catal.*, 1991, **66**, L27-L31.
490. F. Joó and A. Benyei, *J. Organomet. Chem.*, 1989, **363**, C19-C21.
491. A. Benyei and F. Joó, *J. Mol. Catal.*, 1990, **58**, 151-163.
492. A. A. Andriollo, J. Carrasquel, J. Marino, F. A. Lopez, D. E. Paez, I. Rojas, and N. Valencia, *J. Mol. Catal. A: Chem.*, 1997, **116**, 157-165.

493. R. A. Sanchez-Delgado, M. Medina, F. Lopez-Linares and A. Fuentes, *J. Mol. Catal. A: Chem.*, 1997, **116**, 167-177.
494. S. Kolaric and V. Šunjic, *J. Mol. Catal. A: Chem.*, 1996, **110**, 189-193.
495. M. Hernandez and Ph. Kalck, *J. Mol. Catal. A: Chem.*, 1997, **116**, 131-146.
496. M. Hernandez and Ph. Kalck, *J. Mol. Catal. A: Chem.*, 1997, **116**, 117-130.
497. Y. Dror and J. Manassen, *J. Mol. Catal.*, 1977, **2**, 219-222.
498. A. Andriollo, A. Bolivar, F. A. Lopez and D. E. Paez, *Inorg. Chim. Acta*, 1995, **238**, 187- 192.
499. Y. Moroi, "Micelles: Theoretical and Applied Aspects" Plenum, New York, 1992, (a) p.8-15; (b) p.44-47.
500. A. Kumar, G. Oehme, J. P. Roque, M. Schwarze and R. Selke, *Angew. Chem.*, 1994, **106**, 2272; *Angew. Chem. Int. Ed. Engl.*, 1994, **33**, 2197-2199.
501. T. Malmström and C. Andersson, *J. Mol. Catal. A: Chem.*, 1997, **116**, 237-245.
502. F. Gassner and W. Leitner, *J. Chem. Soc., Chem. Commun.*, 1993, 1465-1466.
503. W. Leitner, *Angew. Chem.*, 1995, **107**, 2391-2405; *Angew. Chem. Int. Ed. Engl.*, 1995, **34**, 2207-2221.
504. C. Mercier, G. Mignani, M. Aufrand, and G. Allmang, *Tetrahedron Lett.*, 1991, **32**, 1433- 1436.
505. L. Lavenot, A. Roucoux, and H. Patin, *J. Mol. Catal. A: Chem.*, 1997, **118**, 153-159.
506. G. Mignani, D. Morell, and Y. Colleuille, *Tetrahedron Lett.*, 1985, **26**, 6337-6340.
507. J. Kiji, T. Okano, and T. Hasegawa, *J. Mol. Catal. A: Chem.*, 1995, **97**, 73-77.
508. T. Jeffery, *Tetrahedron Lett.*, 1994, **35**, 3051-3054.
509. A. L. Casalnuovo and J. C. Calabrese, *J. Am. Chem. Soc.*, 1990, **112**, 4324-4330.
510. J. P. Genet, E. Blart, and M. Savignac, *Synlett.*, 1992, 715-717.
511. D. B. Grotjahn and X. Zhang, in Ref. 70, p.123-125.
512. S. Lemaire-Audoire, M. Savignac, C. Dupuis, and J. P. Genet, *Tetrahedron Lett.*, 1996, **37**, 2003-2006.
513. D. B. Grotjahn and X. Zhang, *J. Mol. Catal. A: Chem.*, 1997, **116**, 99-107.
514. J. P. Genet, A. Linquist, E. Blart, V. Mouries and M. Savignac, *Tetrahedron Lett.*, 1995, **36**, 1443-1446.
515. A. I. Roshchin, N. A. Bumagin, and I. P. Beletskaya, *Tetrahedron Lett.*, 1995, **36**, 125-128.
516. R. Rai, K. B. Aubrecht, and D. B. Collum, *Tetrahedron Lett.*, 1995, **36**, 3111-3114.
517. C. Amatore, E. Blart, J. P. Genet, A. Jutand, S. Lemaire-Audoire, and M. Savignac, *J. Org. Chem.*, 1995, **60**, 6829-6839.
518. E. Kuntz, Rhône-Poulenc Ind., *FR 2 366 237*, 1976.
519. E. Monflier, P. Bourdauducq, J. L. Couturier, J. P. Kervennal, and A. Mortreux, *J. Mol. Catal. A: Chem.*, 1995, **97**, 29-33.
520. E. Monflier, P. Bourdauducq, J. L. Couturier, J. P. Kervennal, and A. Mortreux, *Appl. Catal. A*, 1995, **131**, 167-178.
521. E. Monflier, P. Bourdauducq, J. L. Couturier, J. P. Kervennal, I. Suisse, and A. Mortreux, *Catal. Lett.*, 1995, **34**, 201-212.
522. T. Prinz, W. Keim, and B. Driessen-Hölscher, *Angew. Chem.*, 1996, **108**, 1835-1836; *Angew. Chem. Int. Ed. Engl.*, 1996, **35**, 1708-1710.
523. U. F. Schuchardt and M. Zoca, Universidade Estadual de Campinas, Deten Quimica S.A., *BR 92 00 078*, 1993; *Chem. Abstr.*, 1994, **121**, 180427t.
524. R. E. Murray, Union Carbide Corp., *US 4 689 437*, 1987; *Chem. Abstr.*, 1988, **108**, 6646k.
525. W. Baidossi, N. Goren, J. Blum, H. Schumann, and H. Hemling, *J. Mol. Catal.*, 1993, **85**, 153-162.
526. B. M. Novak and R. H. Grubbs, *J. Am. Chem. Soc.*, 1988, **110**, 7542-7543.
527. M. B. France, R. A. Paciello, and R. H. Grubbs, *Macromolecules*, 1993, **26**, 4739-4741.
528. R. H. Grubbs, in Ref. 70, p. 15-22.
529. M. C. Schuster, K. H. Mortell, A. D. Hegeman, and L. L. Kiessling, *J. Mol. Catal. A: Chem.*, 1997, **116**, 209-216.

530. T. I. Wallow and B. M. Novak, *J. Am. Chem. Soc.*, 1991, **113**, 7411-7412.
531. E. Blart, J. P. Genet, M. Safi, M. Savignac, and D. Sinou, *Tetrahedron*, 1994, **50**, 505-514.
532. M. Safi and D. Sinou, *Tetrahedron Lett.*, 1991, **32**, 2025-2028.
533. S. Sigismondi and D. Sinou, *J. Mol. Catal. A: Chem.*, 1997, **116**, 289-296.
534. J. M. Grosselin, H. Kempf, and J. P. Lecoure, Rhône-Poulenc, *EP 90/9887*, 1990.
535. S. Sigismondi, D. Sinou, M. Perez, M. Moreno-Manas, R. Pleixats, and M. Villarroya, *Tetrahedron Lett.*, 1994, **35**, 7085-7088.
536. J. P. Genet, E. Blart, M. Savignac, S. Lemeune, and J. M. Paris, *Tetrahedron Lett.*, 1993, **34**, 4189-4192.
537. S. Lemaire-Audoire, M. Savignac, E. Blart, G. Pourcelot, and J. P. Genet, *Tetrahedron Lett.*, 1994, **35**, 8783-8786.
538. J. P. Genet, E. Blart, M. Savignac, S. Lemeune, S. Lemaire-Audoire, J. M. Paris, and J. M. Bernard, *Tetrahedron*, 1994, **50**, 497-503.
539. S. Sigismondi and D. Sinou, *J. Chem. Res., Synop.*, 1996, 46-47.
540. S. Lemaire-Audoire, M. Savignac, G. Pourcelot, J. P. Genet, and J. M. Bernard, *J. Mol. Catal. A: Chem.*, 1997, **116**, 247-258.
541. E. Kuntz, Rhône-Poulenc Ind., *FR 2 338 253*, 1976.
542. M. Huser and R. Perron, Rhône-Poulenc Chimie, *EP 650 959*, 1995; *Chem. Abstr.*, 1995, **123**, 229261s.
543. M. Bressan, in Ref. 70, p. 173-175.
544. M. Harustiak, M. Hronek, J. Ilavsky, and S. Witek, *Catal. Lett.*, 1988, **1**, 391-394.
545. E. Monflier, E. Blouet, Y. Barbaux, and A. Mortreux, *Angew. Chem.*, 1994, **106**, 2183-2185; *Angew. Chem. Int. Ed. Engl.*, 1994, **33**, 2100-2102.
546. J. W. Francis and P. M. Henry, *J. Mol. Catal. A: Chem.*, 1995, **99**, 77-86.
547. A. M. Tafesh and M. Beller, *Tetrahedron Lett.*, 1995, **36**, 9305-9308.
548. M. Beller, A. Tafesh, C. Kohlpaintner, and C. Naumann, Hoechst AG, *EP 728 734*, 1996; *Chem. Abstr.*, 1996, **125**, 221348w.
549. F. Ragaini and S. Cenini, *J. Mol. Catal. A: Chem.*, 1996, **105**, 145-148.
550. S. Cenini and F. Ragaini, in "Catalytic Reductive Carbonylation of Organic Nitro Compounds", ed. R. Ugo and B. R. James, Catalysis by Metal Complexes, Kluwer, Dordrecht, 1997, vol.20, ch.4; (a) p.152-154; (b) p.156.
551. J. W. Francis and P. M. Henry, *J. Mol. Catal. A: Chem.*, 1996, **112**, 317-326.
552. R. Jira, in Ref. 31, vol.1, ch.2, p.374-393.
553. H. Sertchook, D. Avnir, J. Blum, F. Joó, A. Kathó, H. Schumann, R. Weimann, and S. Wernik, *J. Mol. Catal. A: Chem.*, 1996, **108**, 153-160.
554. H. Schumann, V. Ravindar, L. Meltser, W. Baidossi, Y. Sasson, and J. Blum, *J. Mol. Catal. A: Chem.*, 1997, **118**, 55-61.
555. A. C. Benyei, S. Lehel, and F. Joó, *J. Mol. Catal. A: Chem.*, 1997, **116**, 349-354.
556. J. Bakos, A. Orosz, S. Cserepi, I. Toth, and D. Sinou, *J. Mol. Catal. A: Chem.*, 1997, **116**, 85- 97.
557. C. Larpent and H. Patin, *Tetrahedron*, 1988, **44**, 6107-6118.
558. A. Benyei, J. N. W. Stafford, A. Kathó, D. J. Darensbourg, and F. Joó, *J. Mol. Catal.*, 1993, **84**, 157-163.
559. C. Larpent, G. Meignan, and H. Patin, *Tetrahedron*, 1990, **46**, 6381-6398.
560. C. Larpent and G. Meignan, *Tetrahedron Lett.*, 1993, **34**, 4331-4334.
561. D. J. Darensbourg, F. Joó, A. Kathó, J. N. W. Stafford, A. Benyei, and J. H. Reibenspies, *Inorg. Chem.*, 1994, **33**, 175-177.
562. P. Heymanns, E. Jägers, A. Seidel, and H. Berwe, Hoechst AG, *DE 43 01 310*, 1994.
563. I. Toth, I. Guo, and B. E. Hanson, *J. Mol. Catal. A: Chem.*, 1997, **116**, 217-229.
564. L. J. Mathias and R. A. Vaidya, *J. Am. Chem. Soc.*, 1986, **108**, 1093-1094.
565. C. M. Starks, C. L. Liotta, and M. Halpern, "Phase-Transfer Catalysis: Fundamentals, Applications and Industrial Perspectives", Chapman & Hall, New York, 1994, p.179-182.
566. E. Monflier, G. Fremy, Y. Castanet and A. Mortreux, *Angew. Chem.*, 1995, **107**, 2450-2452; *Angew. Chem. Int. Ed. Engl.*, 1995, **34**, 2269-2271.

567. E. Monflier, S. Tilloy, G. Fremy, Y. Castanet, and A. Mortreux, *Tetrahedron Lett.*, 1995, **36**, 9481-9484.
568. J. R. Anderson, E. M. Campi, and W. R. Jackson, *Catal. Lett.*, 1991, **9**, 55-58.
569. W. R. Jackson, J. R. Anderson, E. M. Campi, Ciptati, Q. J. McCubbin, and Z. Yang, in Ref. 70, p.187-194.
570. E. Monflier, A. Mortreux and Y. Castanet, CNRS, *WO 96/22267*, 1996.
571. E. Monflier, S. Tilloy, G. Fremy, Y. Barbaux, and A. Mortreux, *Tetrahedron Lett.*, 1995, **36**, 387-388.
572. Y. Yoshimura and Y. Tokitoh, *Petrotech(Tokyo)*, 1993, **16**, 341-345.
573. M. E. Davis, *Chemtech.*, 1992, **22**, 498-502.
574. J. P. Arhancet, M. E. Davis, J. S. Merola, and B. E. Hanson, *Nature (London)*, 1989, **339**, 454-455.
575. J. P. Arhancet, M. E. Davis, J. S. Merola, and B. E. Hanson, *J. Catal.*, 1990, **121**, 327-339.
576. J. P. Arhancet, M. E. Davis, and B. E. Hanson, *J. Catal.*, 1991, **129**, 94-99.
577. J. P. Arhancet, M. E. Davis, and B. E. Hanson, *J. Catal.*, 1991, **129**, 100-105.
578. I. T. Horváth, *Catal. Lett.*, 1990, **6**, 43-48.
579. B. B. Bunn, T. Bartik, B. Bartik, W. R. Bebout, T. E. Glass, and B. E. Hanson, *J. Mol. Catal.*, 1994, **94**, 157-161.
580. I. Toth, I. Guo, and B. E. Hanson, in Ref. 70, p.195-197.
581. G. Fremy, E. Monflier, J. F. Carpentier, Y. Castanet, and A. Mortreux, *J. Catal.*, 1996, **162**, 339-348.
582. Z. Xia, U. Klokner and B. Fell, *Fett/Lipid*, 1996, **98**, 313-321.
583. E. Fache, C. Mercier, N. Pagnier, B. Despeyroux and P. Panster, *J. Mol. Catal.*, 1993, **79**, 117-131.
584. R. A. Sheldon and J. K. Kochi, "Metal Catalyzed Oxidations of Organic Compounds", Academic, New York, 1981.
585. J. A. Arhancet, M. E. Davis, and B. E. Hanson, *Catal. Lett.*, 1991, **11**, 129-136.
586. J. M. Brown and S. G. Davies, *Nature (London)*, 1994, **370**, 418-419.
587. R. A. Sheldon, "Chirotechnology", Dekker, New York, 1993, p.301-302.
588. H. Alper, *Adv. Organomet. Chem.*, 1981, **19**, 183-211.
589. L. Cassar, *Chim. Ind. (Milano)*, 1985, **67**, 256-262.
590. E. S. Gore, *Platinum Met. Rev.*, 1990, **34**, 2-9.
591. V. Grushin and H. Alper, in Ref. 70, p. 81-95.
592. L. Cassar and M. Foa, *J. Organomet. Chem.*, 1973, **51**, 381-393.
593. H. Alper and J. K. Currie, *Tetrahedron Lett.*, 1979, 2665-2666.
594. H. Alper and H. des Abbayes, *J. Organomet. Chem.*, 1977, **134**, C11-C14.
595. D. L. Reger, M. M. Habib and D. J. Fauth, *Tetrahedron Lett.*, 1979, 115-116.
596. L. Cassar and M. Foa, *J. Organomet. Chem.*, 1977, **134**, C15-C16.
597. H. des Abbayes, *New J. Chem.*, 1987, **11**, 535-542.
598. M. Foa and F. Francalanci, *J. Mol. Catal.*, 1987, **41**, 89.
599. H. des Abbayes and A. Buloup, *J. Chem. Soc., Chem. Commun.*, 1978, 1090-1091.
600. H. des Abbayes, A. Buloupand and G. Tanguy, *Organometallics*, 1983, **2**, 1730.
601. L. Cassar and M. Foa, Montedison SpA, *US 128 572*, 1978.
602. T. X. Le and J. S. Merola, *Organometallics*, 1993, **12**, 3798-3799.
603. J. S. Merola, T. L. Husebo, K. E. Matthews, M. A. Franks, R. Pafford and P. Chirik, in Ref. 70, p. 33-45.
604. A. C. Benyei, in Ref. 70, p.159-171.
605. M. M. T. Khan, *Platinum Met. Rev.*, 1991, **35**, 70-82.